Zoology: Understanding the Animal World

Zoology: Understanding the Animal World

Annie Wade

STATES
ACADEMIC PRESS
www.statesacademicpress.com

Zoology: Understanding the Animal World
Annie Wade
ISBN: 978-1-63989-577-9 (Hardback)

STATES
ACADEMIC PRESS

Published by States Academic Press,
109 South 5th Street,
Brooklyn, NY 11249, USA

Cataloging-in-Publication Data

Zoology : understanding the animal world / Annie Wade.
p. cm.
Includes bibliographical references and index.
ISBN 978-1-63989-577-9
1. Zoology. 2. Animals. I. Wade, Annie.
QL45.2 .Z66 2022
590--dc23

For more information regarding States Academic Press and its products, please visit the publisher's website www.statesacademicpress.com

Table of Contents

Preface

The field of study that is concerned with animals is known as zoology. It deals with the investigation into individual animals and their constituent parts - animal populations, entire faunas and their relationships with each other, with plants, and with the nonliving environment. Zoology encompasses all aspects of scientific knowledge about animals, such as embryonic development, evolution, behavior, ecological distribution and classification. It is sub-divided into various disciplines like zoology, animal physiology, cognitive zoology and vertebrate zoology. Zoography focuses on the study of animals and their habitats while animal cognition deals with the mental abilities of non-human animals. This book presents the complex subject of zoology in the most comprehensible and easy to understand language. From theories to research to practical applications, case studies related to all contemporary topics of relevance to this field have been included in it. This book aims to serve as a resource guide for students and experts alike and contribute to the growth of the discipline.

A foreword of all chapters of the book is provided below:

Chapter 1 - Zoology is the field of biology that deals with the scientific study of the behavior, structure, physiology, classification and distribution of animals. Zoology can be classified into branches based on different types of animals studied. This chapter serves as an introduction to zoology and its various branches and their related sub branches; **Chapter 2** - Animal classification is a system of grouping animals based on shared characteristics. Taxonomy is the scientific naming and classification of living organisms using a hierarchical order. They are classified into domain, kingdom, phylum, class, order, sub order, families, genus and species. This chapter provides a detailed overview of animal classification and taxonomy for a better understanding of the subject; **Chapter 3** - Animal anatomy is the study of the various forms and structures of animals. Animal physiology deals with the scientific study of the biological processes that occur for animal life to exist as well as the functioning of the animal body parts. The two subjects are closely linked and provide a better understanding of the animal. This chapter has been carefully written to provide an easy understanding of all aspects related to both animal anatomy and physiology; **Chapter 4** - Reproduction is the biological process by which an organism produces an offspring. It is vital for the continuation of life on Earth. This chapter deals with animal reproduction and elaborates the various concepts such as sexual and asexual reproduction, dimorphism and mimicry in an elaborate manner for the reader's benefit; **Chapter 5** - The branch of biology that deals with the scientific and objective study of animal behavior is called ethology. It focuses on animal behavior under normal and hostile conditions and its relation to evolutionarily adaptive traits. Bunting, deimatic behavior, collective behavior, animal consciousness, animal communication, etc. all fall under its domain. This chapter has been carefully written to provide a better understanding of ethology and its related aspects.

At the end, I would like to thank all the people associated with this book devoting their precious time and providing their valuable contributions to this book. I would also like to express my gratitude to my fellow colleagues who encouraged me throughout the process.

Annie Wade

1

Zoology: An Introduction

Zoology is the field of biology that deals with the scientific study of the behavior, structure, physiology, classification and distribution of animals. Zoology can be classified into branches based on different types of animals studied. This chapter serves as an introduction to zoology and its various branches and their related sub branches.

Zoology, or animal biology, is the field of biology that involves the study of animals. It encompasses all aspects of scientific knowledge about animals, like embryonic development, evolution, behavior, ecological distribution, and classification. Zoology is broken into many branches because there are so many different ways to study animals; it is also broken into branches based on which animals are being studied.

People have been interested in learning about animals since ancient times. The prominent ancient Greek philosopher Aristotle took detailed notes on animal observations, and inspired other scientists for many hundreds of years. Many universities were founded in Europe in the 16th Century, and by the mid-17^{th}Century, divisions were founded in universities that focused entirely on animal research.

In the 19th Century, the microscope became commonplace in scientific research, and this opened up a whole new realm of possibility; now, the cells of animals could be studied at the microscopic level. Another breakthrough in zoology occurred when the naturalist Charles Darwin developed the theory of evolution by natural selection. This theory revolutionized zoology and taxonomy (classification). More recently, the discovery of DNA as life's genetic material led to even more new research and knowledge about the natural world and the evolutionary relationships between animals.

Branches of Zoology

Zoography

Zoography, also called descriptive zoology or zoogeography, is the study of animals and their habitats. It is concerned with the geographic ranges of specific populations of animals, their effects on the ecosystems they live in, and the reasons for a specific spatial distribution of an animal species.

Comparative Anatomy

Comparative anatomy is the study of similarities and differences in the anatomy of different types of animals. Closely related animals like mammals share common bones, even if these bones have been extremely modified in shape. For example, bat wings are essentially very long, thin bones that arehomologous to the ones in human fingers. Studying the similar anatomical structures in related organisms provided evidence for evolution from a common ancestor that was later confirmed by genetics research. Comparative anatomy is still used today, often in paleontology, the study of fossils.

Animal Physiology

Animal physiology is the study of the bodily processes that occur in animals that allow them to maintain homeostasis and survive. Homeostasis is the ability of the body to maintain a relatively constant equilibrium even in a changing environment. One example is the regulation of body temperature in mammals. Humans have a normal body temperature of about 37 °C (98.6 °F), even when the environment they are in is much colder. Animal physiology involves the study of processes like temperature regulation, blood pressure and blood flow, and the release of hormones at specific times in the body.

Ethology

Ethology is the study of animal behavior, usually in regard to in their natural environment as opposed to in a lab. Ethology has roots in the work of Darwin, but emerged as a field in the 1930s. It involves the study of animal learning, cognition, communication, and sexuality, and is related to evolutionary biology andecology. Principles from ethology research are also used in animal training.

Behavioral Ecology

Behavioral ecology emerged from ethology. It is the study of evolution as the basis for animal behavior due to ecological pressures, which are constraints placed upon organisms by their environment. Organisms with traits that are well-suited to their environment have a higher likelihood of surviving and reproducing than those who do not. When the proportion of individuals with favored traits increases over a long period of time, evolution can occur. Behavioral ecologists study animals' competition for resources such as food, territory, and mates and the increased reproductive success that certain traits may give.

Groupings by Animal

Zoology is also broken down into subcategories based on the type of animal being studied. For example, a distinction is made between invertebrate zoology and vertebrate zoology. There are also many specific terms for each type of animal that is studied. Some examples are:

- Mammalogy: The study of mammals. A popular type of mammalogy is primatology, the study of primates.
- Ornithology: The study of birds.
- Herpetology: The study of amphibians and reptiles.
- Ichthyology: The study of fish.
- Entomology: The study of insects. Entomology is itself broken down into many categories

because there are so many types of insects. Some examples of its subcategories are Lepidopterology, the study of butterflies and moths, Myrmecology, the study of ants, and Coleopterology, the study of beetles.

Figure: Coral reefs, are full of biodiversity; many fish and other sea animals live on corals, which are also animals themselves.

Vertebrate Zoology

Vertebrate zoology is the study the biology of fish, amphibians, birds, reptiles and mammals.

Herpetology

Herpetology is the scientific study of amphibians and reptiles. Like most other fields of vertebrate biology (e.g., ichthyology, mammalogy), herpetology is composed of a number of cross-disciplines: behaviour, ecology, physiology, anatomy, paleontology, taxonomy, and others. Most students of recent forms are narrow in their interests, working on only one order or suborder (e.g., frogs, salamanders, snakes, lizards). A paleontologist is more likely to work with both amphibians and reptiles or with intermediate forms.

Herpetology as a unified science apparently stems from the ancient tendency to lump together all creeping animals. Modern herpetology is a truly popular science, in which amateurs have made many valuable contributions in such areas as distribution, behaviour, and even taxonomy. The major part of the more technical research is carried out at universities and museums, as well as in the field.

Research into the biology of different amphibians and reptiles has contributed much to the field of general biology as for example, larval frogs and salamanders in the understanding of embryological concepts, iguanid lizards with the development of the sub-discipline of population ecology, and snake venom in increasing the understanding of human cardiac and neurological disorders.

Ichthyology

Ichthyology is the branch of zoology (the study of animals) that deals with fish. It includes the study of the development, anatomy (structure), physiology (function), behavior, classification, genetics, and ecology of fish, among other things. Since fish are a major food source for people, the study of ichthyology also has economic importance.

Taken from the Greek word ichthys for fish, ichthyology had its beginning with the Greek philosopher Aristotle, since the ancient world was more interested in and more knowledgeable about fish than they were about many other animals. This may have been because fish were both a relatively easy-to-obtain source of food as well as an animal group that was readily accessible, since fishing is one of humankind's oldest occupations. Until the end of the nineteenth century, however, more attention was paid to describing and classifying fish than any other aspect. By then, ichthyology was well on its way to becoming a separate field of zoology. By the end of the first half of the twentieth century, the emergence of oceanography (the science of the ocean) and the newfound ability to conduct underwater observations, allowed scientists to be able to study fish in their natural environment for the first time. The development of improved techniques for keeping fish in tanks for study also spurred further advances.

There are more than 22,000 known species of fish in the world, and they live in nearly every imaginable body of water, from stagnant ponds to the deepest oceans. They live in water all of the time and breathe through gills. Together with mammals, birds, reptiles, and amphibians, fish are one of the major groups of vertebrates (animals with a backbone). They are considered the most successful vertebrate group, outnumbering birds two to one and mammals seven to one. An ichthyologist, therefore, must contend with a great variety of subjects, from the bony, snakelike eel and the shark with all cartilage and no bone, to the bioluminescent deep-sea fish that can make its own light.

Today, fish are not studied just for their own sake or to simply learn more about them. Since fish are a major food source and fishing is an important industry, a great deal of fishery research is conducted in government laboratories as well as institutional aquariums. It is not surprising that much of this work is aimed at learning more about diseases in fish as well as understanding the effect that pollution has on them. Fishes are as vulnerable to infections by viruses as are higher vertebrates, and often they are the first to show signs of disease. They are also susceptible to tumors, and sick fish are a signal that they live in an environmentally degraded body of water. Increasingly, ichthyologists must know as much about the environment of a certain type of fish as they do about the fish itself in order to note any irregularities in the environment and thus in the fish.

Mammalogy

Mammalogy is the branch of biology that deals with the study of mammals. It encompasses such diverse areas as the structure, function, evolutionary history, ethology, taxonomy, management and economics of mammals. Approximately 4,200 species of living mammals and numerous extinct species comprise the material for study. Included are egg-laying echidnas, the platypus, pouched marsupials, tiny shrews, bats, mice, whales, apes and elephants, to name only a few.

The study of mammals can be as diverse as the organisms themselves. A mammalogist might study a wide variety of topics on a particular species or group of mammals or might take a comparative approach and investigate one aspect with regard to a wide variety of mammals. The major subdivisions of the science of mammalogy include the following:

The study of a species' way of life, including descriptions of the habitats in which it lives, the food it eats, the predators that eat it, the stages of its maturation and reproduction and its social structure.

- Taxonomy and Systematics: The study of the classification of mammals into distinct orders, families, genera, species and subspecies, thereby defining the geographic distribution of each taxon (a taxonomic category or group, such as a phylum, order, family, genus or species). In addition, the evolutionary relationships among extinct (paleontology) and/or living taxa are analyzed.
- Anatomy and Physiology: The study of mammalian body structures and tissues and how each functions.
- Ethology: The study of the behavior of an animal and how those behaviors influence its survival and reproduction.
- Ecology: The study of interactions between mammals and their environments (non-living and living). This discipline includes the study of special adaptations to environmental factors (physiological ecology) and the study of interactions within and among species (population and community ecology).
- Management and Control: The interactions between humans and other mammals in which humans manipulate either the environment or populations of mammals to favor the use and/or survival of certain species and to regulate or even reduce the populations of species whose activities conflict with human interests.

Ornithology

Ornithology is the branch of zoology that deals with the systematic study of birds, including their spindle-shaped body, well-developed flight muscles, physical appearance, habitats, along with their migration patterns and lot more.

There are approximately 9,000 to 10,000 living species of bird existing on this planet and altogether, 25 different orders of bird including fowl, barbets, doves, raptors woodpeckers, parrots, and lot more. All these bird species are classified into the Kingdom Animalia, Class: Aves and Phylum: Chordata.

Scientists who study the anatomy and physiology of birds are called Ornithologists. There are many scientists, who have contributed their life in the research and discoveries of birds and their evolution. Charles Darwin, Richard Jefferies, Thomas Bewick, Gilbert white are a few among the famous Ornithologists.

Significance of Ornithology

Ornithology mainly refers to natural science which mainly deals with the study of birds, their ecology, evolution, physiology and more about their preferred habitats. The science of ornithology has a long history and has helped in the development of several key concepts in evolution, the definition of species, behaviour and the process of speciation, ecological niches, Biodiversity conservation, instinct, phylogeography and island biogeography. Ornithology explains to us about the unique characteristics of birds, which includes:

- Migration: There are several species of birds that migrate long ranges and to a different place in search of food, shelter and also in search of warm or cold conditions and also in search of best environmental specifications.
- Mating Behaviors: There are many species of birds that mate for only one season, while others mate multiple times during one season.
- Vocal call or a song: There are few species of birds, which produces a type of vocal call or a song, which is used for a variety of purposes including:
    - Warn predators.
    - Attracting its mate.
    - Establishing territory.
    - Co-ordinating flock movements.
    - To identify their own offspring.
    - Sexual and Species identification.
    - Some practice, for their perfection of the song.
    - Some birds also sing-song to mimic or imitate.

According to the research or the records found, there are two main categories of birds:

Archaeornithes

This group of birds are extinct. They had a toothed beak, with a long lizard-like tail. E.g. Archaeopteryx.

Neornithes

This type of bird includes both extant as well as extinct birds. These bird species have no teeth and a short tail. For example Penguin, Grey Heron, Kingfisher, Duck, etc. This was a brief introduction to ornithology. It is a vast field and is classified into three main groups, based on its modes of researches:

- Collection,
- Field-based research,
- Laboratory-based research.

Batrachology

Batrachology is the study of amphibians. Most people use the term "herpetology" but this includes reptiles. Evolutionary biologists all agree that taxonomy should reflect natural groups, i.e., monophyletic clades. So, why do we still use a grossly "polyphyletic" concept of herpetology as including the Amphibia? The only seemingly rational argument, besides the worn out appeal to "tradition," is that similar logistics and field protocols are involved in studying extant amphibians and reptiles. If this argument holds water, then I should like to be the first to propose the new science of "flying tetrapods," since the study of both bats and birds utilizes mist nets, gloves and cloth bags. Otherwise, "herpetology" should be restricted to reptiles.

Amphibia is a major class of vertebrate included under phylum Chordata. Amphibia refers to "double life", or life in water and on land. Class Amphibia includes the salamanders, frogs, toads, and caecilians. All amphibians have very moist skin and produce shell less eggs. They are therefore incredibly dependent on water for reproduction. Amphibians are ectothermic and have permeable skin; they rely upon their external environment for regulating body temperature and moisture loss. They are both gill and lung breathers usually gills in the larval stage, replaced by lungs in the adult.

They inhabit a variety of life zones. Except oceans, they can be found from deserts to the sub-polar region, from sea level to snow line, every imaginable type of freshwater, from the ground up to the highest treetop.

There are 3 orders of amphibians Apoda, Urodela and Anura. Apoda are limbless amphibians. Order Urodela are characterized by having a true tail. They include salamanders and Newts. Order Anura is comprised of the frogs and toads. The distinguishable characteristic of this order is the absence of a tail in the adult form and the presence of tympanum. Order Anurans are the most diverse and widespread of the 3-extant amphibian order.

Amphibians are referred to as an indicator species because they are particularly sensitive to changes in their environment and monitoring their status is a way to assess ecosystem health. Amphibians play a very important role in the food chain of both terrestrial and aquatic ecosystems. The general ecological importance of amphibians lies in them being predators acting as primary and secondary carnivores on insects, some of which are crop pests or disease vectors. Amphibians are valuable for natural biological pest control.

There are more than 5100 extant amphibian species have been described and the number increases at a yearly rate of approximately 70 to 100 newly discovered species. Amphibians seem to have once again reached a level of diversity comparable to their first "golden age", the Carboniferus and the Permian. With approximately 4500 valid species, Anurans are by far the most species-rich amphibian group. India houses 314 amphibian species out of which 279 belong to order Anura. About 138 species of amphibians are endemic to India. Odisha is home to 26 amphibian species. Only anuran amphibia have so far been reported in Odisha.

Invertebrate Zoology

Invertebrate zoology is therefore the scientific study of the group of animals known as invertebrates.

Helminthology

Helminthology is the study of parasitic worms. This field deals with the study of their taxonomy and the effect on their hosts. Helminths are a large, multicellular organism that is visible to the eye once in the adult stage of their life cycle. Helminths and multicellular eukaryotes can either be free-living or parasitic.

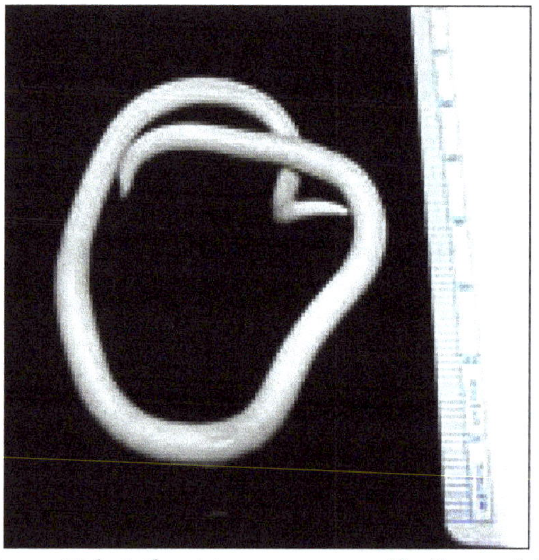

Figure: Helminths: A tapeworm, specifically an adult Ascaris lumbriocoides worm.

In their adult form, helminths are unable to multiply in humans and utilize numerous mechanisms of transmission to ensure reproductive success. These parasites live in and feed on hosts which allow them to obtain nourishment while disrupting the hosts' nutrient absorption. Parasitic worms are commonly found within the intestine and thus, are called intestinal parasites. They are able to live in both humans and animals. The major groups of parasitic helminths include:

- Platyhelminths (flatworms),
- Acanthocephalins (thorny-headed worms),
- Cestodes (tapeworms),

- Trematodes (flukes),
- Nematodes (roundworms).

The classification and identification of helminths are dependent on numerous factors including body shape, body cavity, body covering, digestive tubing, sex and type of attachment organs. Platyhelminths (flatworms) include both trematodes (flukes) and cestodes (tapeworms). Specifically, tapeworms are characterized using the above criteria and are organized in a segmented plane. They lack a body cavity and have a *tegument*body covering. Tapeworms lack a digestive tube and are hermaphroditic. They utilize suckers or *bothridia*, and *rostellum* with hooks for an attachment organ.

Trematodes are characterized by an unsegmented plane for body shape. They also lack a body cavity and have a tegument for body covering. However, the digestive tube for trematodes ends in the cecum. Trematodes are hermaphroditic and utilize oral suckers, ventral suckers or *acetabulum* for attachment organs.

Nematodes are characterized by a cylindrical body shape and do indeed have a body cavity. Its body covering is a cuticle and the digestive tube ends in the anus. The sex of nematodes is *dioecious* (distinct male and female organisms). Lastly, their attachment organs range from lips, teeth, filariform extremities and dentary plates.

Malacology

Malacology is the study of mollusks. The phylum Mollusca includes such soft-bodied invertebrates such as snails, clams, oysters, and squid. Most seashells were once living mollusks. The field of Malacology is not a very well known branch of science, yet it plays an important role in biological ecosystems all over the world. Studying mollusks helps scientists understand some of the most complicated problems of this century, such as cellular cancer and pollution control.

Mollusks provide a vast array of uses and benefits. They are an enormous source of food for both humans and animals; their shells can be used to make tools and implements, and are also used as a calcium supplement. Bivalves filter out toxic waste out of the water, and Cone Shell venom is used as medical serum. There is even a snail whose large cellular structure is being used to test cures for cancer. Here's a list of the uses for mollusks:

- Food: Escargot, calamari, shellfish, etc.
- Tools/weapons: Tough shells make rudimentary implements.
- Roadwork: Crushed shell used to pave streets.
- Decoration: Abalone shell (mother of pearl), seashells, pearls, etc.
- Clothing: Mussel shell buttons, purple dye (Murex sp).
- Medical- Calcium supplements, disease and cancer research.
- Pest control: Carnivorous snails prey on harmful pest species.
- Pollution control: Bivalves filter toxins from water.
- Horticulture: Many snails and slugs recycle dead plants to create a natural fertilizer.

- Aquarium trade: Aquatic snails look pretty and eat algae off the glass.
- Religion: Jewish culture uses Murex trunculus to create a special, sacred blue dye for tassels.
- Currency: Some shells used as money in certain countries.

Classifying Mollusks

The Phylum Mollusca

All mollusks are classified under the phylum Mollusca. There are six classes of mollusks:

Class Cephalopoda

This class includes the highly intellegent, many-armed orders of mollusks. These are only found in marine environments. Each animal has at least eight arms, and some have two additionial tentacles. Squids have a rigid internal structure called a pen, and cuttlefish have an internal shell called a cuttlebone. Nautili have an external shell. Octopi have neither. All cephalopods have a parrot-like beak with which they dismantle their prey (squid will bite if handled). These animals are so intelligent, they are used by scientists to study how the brain learns and works:

- Octopus,
- Squid,
- Cuttlefish,
- Nautilus,
- Argonaut,
- Spirula.

Class Gastropoda

This is the largest class of mollusks, and includes those with either one or no external shell on the dorsal side:

- Land/Freshwater snails,
- Slugs,
- Marine snails,
- Pteropods,
- Nudibranchs.

Class Monoplacophora

This class contains gastroverms, mollusks with a single dorsal shell and an esposed ventral side. These live on rocks in the deep sea. If one senses danger, it will seal itself to a rock to protect itself, often requiring many hundreds of pounds of pressure to remove.

- Limpets,
- Abalones.

Class Pelecypoda

This class encompasses the two-halved, hinged shell variety of mollusks known as Bivalves. These are strictly aquatic:

- Oysters,
- Mussels,
- Clams,
- Scallops,
- Cockles.

Class Polyplacophora

This class includes mollusks with jointed dorsal armor which live on tidal rocks. Each animal has eight armored plates on its back which protect it from predators. In the event of an attack, it clings tightly to the rock face:

- Chitons.

Class Scaphapoda

This class includes the small, tooth-shaped mollusks which live buried under mud in the ocean. Some species live at depths of over 15,800 feet:

- Tusk Shells,
- Deep-sea Cadulus.

Classifying Landsnails

Land snails, which are what Molluskman.com is primarily focused on, are categorized as Pulmonates under the class Gastropoda. There are three orders of gastropods:

- Opisthobranchia: These mollusks are marine sea slugs, most of which lack external shells. Some of these animals have bright colors and stinging cells:
    - Nudibranchs,
    - Sea Hares,
    - Bubble Shells,
    - Pteropods (sea butterflies).
- Prosobranchia: This is the order involving marine snails. These mollusks are prized by collectors for their beautiful shells. Most marine snails have an operculum.

  - Whelks,
  - Conchs,
  - Tritons,
  - Volutes,
  - Augers,
  - Cowries,
  - Murex.
- Pulmonata: Pulmonates are distinguished by the fact that their mantle cavity can be used as a lung. Order Pulmonata includes aquatic and terrestrial snails and slugs:
  - Land Snails,
  - Pond Snails,
  - Slugs.

Anatomical Terms

- Aperture: The opening of a univalve shell.
- Apex: The top or peak of the dorsal side of a univalve shell.
- Comnunella: The spiraled column that runs up the center of shells, kind of like the center column of a spiral staircase.
- Mantle: This part of the body differs in function for each type of mollusk: in gastropods, it is a sheet-like organ that covers the visceral mass, contains organs, and secretes calcium onto the shell; in bivalves, it is most recognizable as the part that is eaten; in cephalopods, it is the sac or tube that contains internal organs and serves as a cranium.
- Operculum: A tough, corneous, trapdoor-like object that covers a shell's aperture and protects the animal when it retracts into its shell. Found only in some gastropods.
- Periostracum: A thin, film-like layer that covers the surface of many shells, and gives some their color. It is comparable to the periosteum that covers bones.
- Radula: A tongue-like organ with thousands of tiny teeth which certain mollusks use to rasp away at their food. It's like a belt with teeth, or a cheese grater.
- Umbilicus: The hole that is formed by the inside of the comnunella that runs up the center of a univalve shell, open at the anterior end and ending at the posterior, or apex.
- Whorl: The twisting form of a gastropod shell; the number of whorls depends on how many times the shell twists around in a complete circle.

Cnidariology

Cnidariology is a sub-discipline of invertebrate zoology that studies the cnidarians. Invertebrate zoology is concerned with the invertebrates. Invertebrates are animals that lack a vertebral column (a backbone). Since there are a number of invertebrate species there are taxonomic-based sub-disciplines of invertebrate zoology, particularly arthropodology (the study of arthropods), cnidariology, and helminthology (the study of helminths).

Cnidariology deals with studying the cnidarians. Cnidarians are those belonging to phylum Cnidaria. They include the jellyfishes, sea anemones,corals, hydras, *etc.* They are characterized primarily by possessing cnidae, i.e. specialized cellular structures that fire off thread-like tubules producing a stinging effect.

An expert on cnidariology is called *cnidariologist.* A cnidariologist would be concerned and responsible in conducting research in cnidarians. The research may be associated with the identification and classification of cnidarian species. It may also be about their distribution, conservation, and ecological relationship of the cnidarians to other living things and their environment. The research may also encompass other related fields such as genetics, evolution, and medical importance of cnidarians.

Arthropodology

Arthropodology is a biological discipline that focuses on the arthropods. Arthropods are invertebrates that belong to the phylum Arthropoda. This phylum is the largest among the phyla in the animal kingdom. They include the insects, the arachnids, the myriapods and the crustaceans. These animals are grouped under this phylum based on the features common to them. They have multiple paired jointed limbs and a hard chitinous exoskeleton. The exoskeleton is a hard outer casing that protects their soft internal organs. Arthropodology deals with these animals and aims to understand their evolution, anatomy, biology, and ecological importance.

Arthropodology that studies the parasitic mode of life of certain arthropods is referred to as medical arthropodology. Apart from parasitic arthropods, there are also arthropod species that act as vectors of disease. An example of that would be the mosquito. Certain species of mosquitoes serve as vectors for disease-causing viruses and parasites. Mosquito-borne viral diseases include the yellow fever, dengue fever, and zika. Mosquito-borne parasitic diseases are malaria and lymphatic filariasis.

Arthropodology may be subdivided into the following fields based on the animal groups that they study: arachnology (studies spiders and other arachnids), entomology (studies insects), carcinology (studies crustaceans), and myriapodology (studies myriapods).

Arachnology

Arachnology is the study of arachnids. Although mites and ticks are arachnids, for a number of reasons they have their own science, called Acarology, devoted to them.

Arachnids are an ancient group of creatures dating back to the Silurian period, more than 420 million years ago (earlier ones have allegedly been found but there status as Arachnids is arguable). They belong to the Arthropods ('jointed limbs') that account for over 80% of all living animals and include the insects, crustaceans, millipedes, and centipedes. Arthropods are characterised by their tough segmented outer shell, or exoskeleton, and jointed limbs.

Arachnids have four pairs of legs, a pair of jaws (the *chelicerae*) and a second pair of appendages (the *pedipalps*). They do not have antennae or wings. Included in this diverse group are:

- Spiders,
- Scorpions,
- False-scorpions,
- Harvestmen or 'daddy-long-legs',
- Whip scorpions,
- Palpigrades,
- Whip spiders,
- Ricinuleids,
- Sun and wind spiders,
- Mites.

Arachnids have a body usually divided into two parts, the prosoma (*cephalothorax* or *prosoma*) and *opisthosoma* ('abdomen'), this contrasts with insects which have their body divided into three parts: head, thorax and abdomen. Mites and harvestmen bodies are less obviously divided, having the prosoma and opisthosoma fused together. The prosoma carries six pairs of appendages: the four pairs of legs, *pedipalps* and chelicerae. Pedipalps and chelicerae are modified in the different arachnid groups and may form spiny grasping limbs or lobster-like claws. The prosoma may also carry varying numbers of eyes.

Virtually all arachnids are terrestrial. However, some spiders, mites and palpigrades occur in both fresh water and intertidally on the seashore in shallow salt water. Halacarid mites are known from 4000m depth seawater. Breathing is through the use of *tracheae* (breathing tubes like those found in insects) lung-books or, as in spiders, both.

The vast majority of arachnids are carnivorous, active predators. Some are scavengers and others parasites. The mites include groups that are herbivores, carnivores, parasites (on both plants and animals) and detritivores (feeding on dead plants and animals). Scorpions and false scorpions have pedipalps which have pincers at their ends which are used, amongst other things, to catch, hold and even kill prey. In most arachnids digestion occurs outside the body, digestive juices being poured into or over their food that is chewed and held by the jaw-like chelicerae.

If you are a predator, reproduction is often a risky business as partners can be potential meals. Arachnids have developed 'indirect sperm transfer' as a way of minimising the risk of being eaten. With this form of mating, there is no intimate coupling so males have a better chance of survival. Some males avoid the risk altogether by depositing packets of sperm (*spermatophores*) for the female to pick up, either in her presence or randomly. To 'neutralise' the aggressive nature of females, males have adopted many techniques. They may grasp the pedipalps of the female so she cannot attack, offer her an insect to eat while he mates, tie her down, seduce her with vibrations, or enact a complex dance to signal who he is and what he intends. Eggs may be laid or carried around by the female or, as in scorpions; the young may be born live. The young may grow up independently or be cared for by their mother until old enough to fend for themselves. Some mothers even feed a milky secretion to their young. The use of silk and silk-related compounds is widespread in the group, particularly in the spiders. It is used in mating, nesting, hiding, hunting, trapping and even ballooning.

Entomology

Entomology is a branch of biology dealing with the study of insects. It includes morphology, physiology, behavior, genetics, biomechanics, taxonomy, ecology, etc. of insects. Any scientific study that focuses on insects is considered an entomological study. The scientific study of insects is believed to have formally occurred in the 16th century. The person that scientifically studies insects is called entomologist.

Insects are a group of arthropods characterized by having a pair of antennae, three pairs of mouth organs, spiracles and tracheae involved in respiration, etc. These organisms comprise the class Insecta of phylum Arthropoda. Examples of such organisms are bees, wasps, ants, beetles, butterflies, moths, dragonflies, fireflies, flies, termites, grasshoppers, crickets, etc. Since this group of organisms comprises a large number of species, entomology has been divided into many subfields based on the kind of insects under study. These subfields are as follows:

- Coleopterology: Beetles,
- Dipterology: Flies,
- Hemipterology: True bugs,
- Isopterology: Termites,
- Lepidopterology: Moths and butterflies,
- Melittology (or Apiology): Bees,
- Myrmecology: Ants,
- Orthopterology: Grasshoppers, crickets, etc,
- Trichopterology: Caddis flies,
- Vespology: Social wasps.

Carcinology

Carcinology is a branch of zoology that deals with insects. It is particularly one of the sub-fields of arthropodology. Arthropodology, in particular, is the biological science that studies arthropods. Arthropods are invertebrates (of the phylum Arthropoda) that are characterized mainly by possessing a hard chitinous exoskeleton and multiple paired jointed limbs. Since arthropodology involves the largest phylum of the animal kingdom it is further subdivided into these sub-disciplines: arachnology (the study of arachnids), entomology (the study of insects), carcinology (the study of crustaceans), and myriapodology (the study of myriapods).

Carcinology is primarily concerned with the studying of crustaceans. It attempts to study and understand various biological aspects of crustaceans, such as development, taxonomy, anatomy, physiology, genetics, evolution, distribution, and ecology of crustaceans. It has taxon-based subdivisions. An example is astacology. It is the study of crayfish. Crayfish are freshwater crustaceans characterized by their feather-like gills and morphological features similar to a lobster. Another sub-discipline is cirripedology, which is the study of barnacles. Copepodology is the sub-discipline that studies copepods.

An expert in this field is called a *carcinologist*. The carcinologist performs studies or research that are focused at understanding the biology of crustaceans as well as their identification and classification, evolutionary relationships, ecological relationships, and distribution.

Myriapodology

Myriapodology is a branch of zoology that deals with myriapods. It is one of the sub-fields of arthropodology. Arthropodology, in particular, is the biological science that studies arthropods. Arthropods are invertebrates (of the phylum Arthropoda) that are characterized mainly by possessing a hard chitinous exoskeleton and multiple paired jointed limbs. Since arthropodology involves the largest phylum of the animal kingdom it is further subdivided into these sub-disciplines: arachnology (the study of arachnids), entomology (the study of insects), carcinology (the study of crustaceans), and myriapodology (the study of myriapods).

Myriapodology deals primarily with the myriapods with the intent of understanding the myriapods' development, taxonomy, biology, evolution, distribution, and ecology. Myriapods include the millipedes, the centipedes, the symphylans, the pauropodans, and the arthropleurideans. An expert in this field is called a myriapodologist. The myriapodologists conduct studies in order to understand the biology of crustaceans as well as their identification and classification, evolutionary relationships, ecological importance, and distribution.

References

- Zoology: biologydictionary.net, Retrieved 11, July 2020
- Herpetology, science: britannica.com, Retrieved 29, March 2020
- Mammalogist: aboutbioscience.org, Retrieved 15, June 2020
- Ornithology, biology: byjus.com, Retrieved 07, February 2020
- Malacology: molluskman.com, Retrieved 14, August 2020
- Cnidariology: biologyonline.com, Retrieved 09, May 2020
- Arthropodology: biologyonline.com, Retrieved 16, January 2020
- Carcinology: biologyonline.com, Retrieved 21, April 2020
- Myriapodology: biologyonline.com, Retrieved 13, July 2020

2

Animal Classification and Taxonomy

Animal classification is a system of grouping animals based on shared characteristics. Taxonomy is the scientific naming and classification of living organisms using a hierarchical order. They are classified into domain, kingdom, phylum, class, order, sub order, families, genus and species. This chapter provides a detailed overview of animal classification and taxonomy for a better understanding of the subject.

Animal Kingdom

Animals are eukaryotic, multicellular, species belonging to the Kingdom Animalia. Every animal has its own unique characteristics. They obtain their energy either by feeding on plants or on other animals. There are millions of species which have been identified; few share similar characteristics while others differ drastically.

Classification of Animal Kingdom

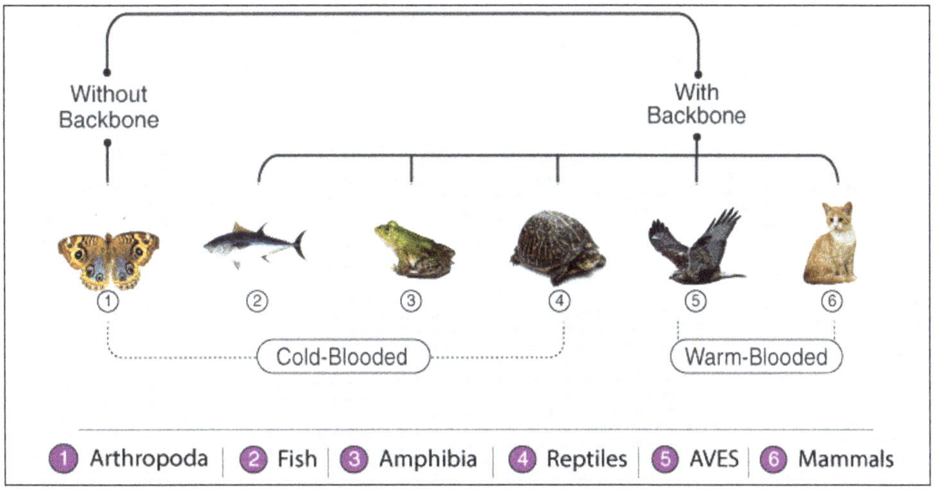

Animals are classified based on their characteristics. They are eminent from algae, plants, and fungus where rigid cell walls are absent. Some are also heterotrophic; in general, they digest their food within the internal chambers which again distinguish them from algae and plants. Another elite character of these species is that they are motile, except in certain life stages.

Features of the Animal Kingdom

The animal kingdom is very diverse, but animals share many common characteristics, such as methods of development and reproduction. Animal evolution began in the ocean over 600 million years ago with tiny creatures that probably do not resemble any living organism today. Since then, animals have evolved into a highly-diverse kingdom. Although over one million extant (currently living) species of animals have been identified, scientists are continually discovering more species as they explore ecosystems around the world. The number of extant species is estimated to be between 3 and 30 million.

But what is an animal? While we can easily identify dogs, birds, fish, spiders, and worms as animals, other organisms, such as corals and sponges, are not as easy to classify. Animals vary in complexity, from sea sponges to crickets to chimpanzees, and scientists are faced with the difficult task of classifying them within a unified system. They must identify traits that are common to all animals as well as traits that can be used to distinguish among related groups of animals. The animal classification system characterizes animals based on their anatomy, morphology, evolutionary history, features of embryological development, and genetic makeup. This classification scheme is constantly developing as new information about species arises. Understanding and classifying the great variety of living species help us better understand how to conserve the diversity of life on earth.

Even though members of the animal kingdom are incredibly diverse, most animals share certain features that distinguish them from organisms in other kingdoms. All animals are eukaryotic, multicellular organisms, and almost all animals have a complex tissue structure with differentiated and specialized tissues. Most animals are motile, at least during certain life stages. All animals require a source of food and are, therefore, heterotrophic: ingesting other living or dead organisms. This feature distinguishes them from autotrophic organisms, such as most plants, which synthesize their own nutrients through photosynthesis. As heterotrophs, animals may be carnivores, herbivores, omnivores, or parasites. Most animals reproduce sexually with the offspring passing through a series of developmental stages that establish a fixed body plan. The body plan refers to the morphology of an animal, determined by developmental cues.

(a) (b)

Heterotrophs: All animals are heterotrophs that derive energy from food. The (a) black bear is an omnivore, eating both plants and animals. The (b) heartworm Dirofilaria immitis is a parasite that derives energy from its hosts. It spends its larval stage in mosquitoes and its adult stage infesting the heart of dogs and other mammals.

Complex Tissue Structure

Animals, besides Parazoa (sponges), are characterized by specialized tissues such as muscle, nerve, connective, and epithelial tissues. As multicellular organisms, animals differ from plants and fungi because their cells don't have cell walls; their cells may be embedded in an extracellular matrix (such as bone, skin, or connective tissue); and their cells have unique structures for intercellular communication (such as gap junctions). In addition, animals possess unique tissues, absent in fungi and plants, which allow coordination (nerve tissue) and motility (muscle tissue). Animals are also characterized by specialized connective tissues that provide structural support for cells and organs. This connective tissue constitutes the extracellular surroundings of cells and is made up of organic and inorganic materials. In vertebrates, bone tissue is a type of connective tissue that supports the entire body structure. The complex bodies and activities of vertebrates demand such supportive tissues. Epithelial tissues cover, line, protect, and secrete; these tissues include the epidermis of the integument: the lining of the digestive tract and trachea. They also make up the ducts of the liver and glands of advanced animals.

The animal kingdom is divided into Parazoa (sponges) and Eumetazoa (all other animals). As very simple animals, the organisms in group Parazoa ("beside animal") do not contain true specialized tissues. Although they do possess specialized cells that perform different functions, those cells are not organized into tissues. These organisms are considered animals since they lack the ability to make their own food. Animals with true tissues are in the group Eumetazoa ("true animals"). When we think of animals, we usually think of Eumetazoans, since most animals fall into this category.

Sponges: Sponges, such as those in the Caribbean Sea, are classified as Parazoans because they are very simple animals that do not contain true specialized tissues.

The different types of tissues in true animals are responsible for carrying out specific functions for the organism. This differentiation and specialization of tissues is part of what allows for such incredible animal diversity. For example, the evolution of nerve tissues and muscle tissues has resulted in animals' unique ability to rapidly sense and respond to changes in their environment. This allows animals to survive in environments where they must compete with other species to meet their nutritional demands.

Basis for Animal Kingdom Classification

Classification of Animal Kingdom is based on various fundamental features like:

- Levels of Organisation,
- Symmetry,
- Diploblastic and Triploblastic Organisation,
- Coelom development,
- Segmentation of the body,
- Presense or absence of Notochord.

The broad classification of Animalia based on common fundamental features:

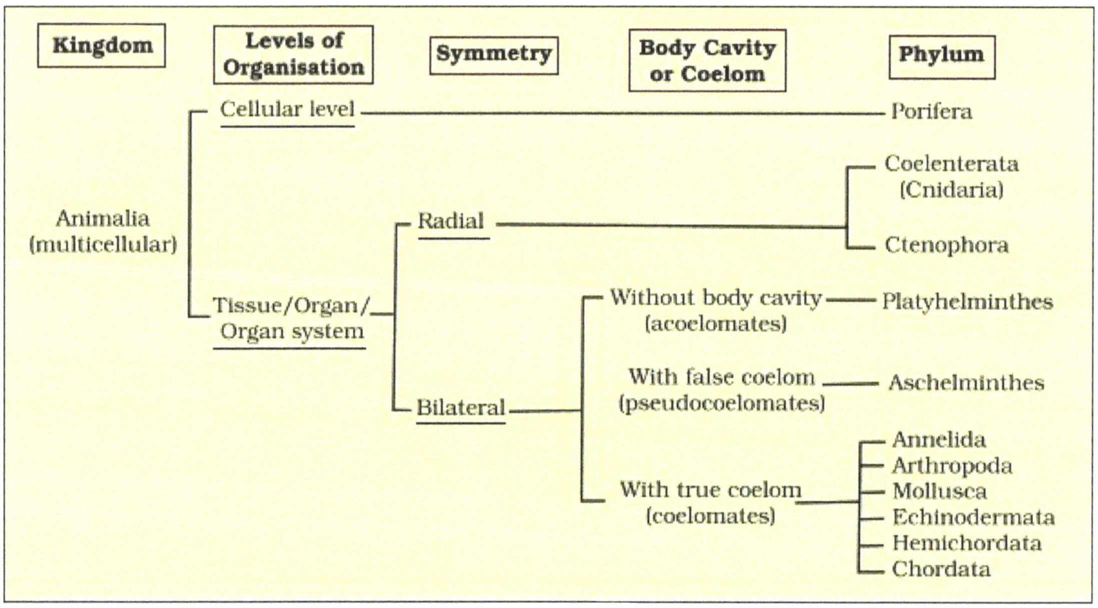

Levels of Organisation

- Though all members of Animalia are multicellular, all of them do not exhibit the same pattern of organisation of cells.
- For example, in sponges, the cells are arranged as loose cell aggregates, i.e., they exhibit cellular level of organisation. Some division of labour (activities) occur among the cells.
- In coelenterates, the arrangement of cells is more complex. Here the cells performing the same function are arranged into tissues, hence is called tissue level of organisation.
- A still higher level of organisation, i.e., organ level [organ level of organisation] is exhibited by members of Platyhelminthes and other higher phyla where tissues are grouped together to form organs, each specialised for a particular function.

- In animals like Annelids, Arthropods, Molluscs, Echinoderms and Chordates, organs have associated to form functional systems, each system concerned with a specific physiological function. This pattern is called organ system level of organisation.
- Organ systems in different groups of animals exhibit various patterns of complexities.
- For example, the digestive system in Platyhelminthes (incomplete digestive system) has only a single opening to the outside of the body that serves as both mouth and anus, and is hence called incomplete. A complete digestive system has two openings, mouth and anus.
- Similarly, the circulatory system may be of two types: open type in which the blood is pumped out of the heart and the cells and tissues are directly bathed in it and closed type in which the blood is circulated through a series of vessels of varying diameters (arteries, veins and capillaries).

Symmetry

- Animals can be categorised on the basis of their symmetry.
- Sponges are mostly asymmetrical, i.e., any plane that passes through the centre does not divide them into equal halves.
- When any plane passing through the central axis of the body divides the organism into two identical halves, it is called radial symmetry. Coelenterates, Ctenophores and Echinoderms have this kind of body plan.
- Animals like Annelids, Arthropods, etc., where the body can be divided into identical left and right halves in only one plane, exhibit bilateral symmetry.

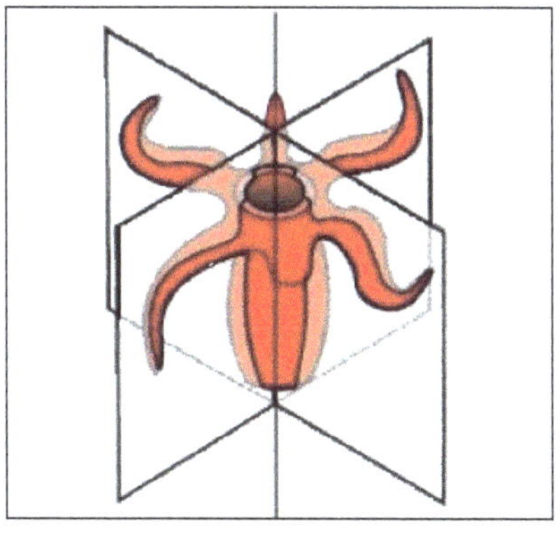

(a) Radial symmetry

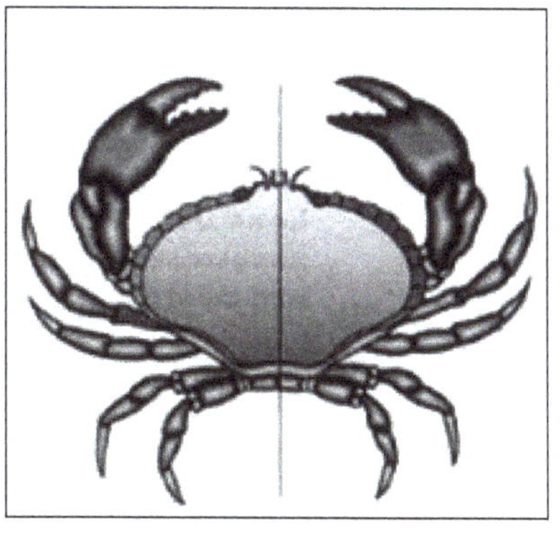

(b) Bilateral symmetry

Diploblastic and Triploblastic Organisation

- Animals in which the cells are arranged in two embryonic layers, an external ectodermand an internal endoderm, are called diploblastic animals, e.g., Coelenterates. An undifferentiated layer, mesoglea, is present in between the ectoderm and the endoderm.
- Those animals in which the developing embryo has a third germinal layer, mesoderm, in between the ectoderm and endoderm, are called triploblastic animals (platyhelminthes to chordates).

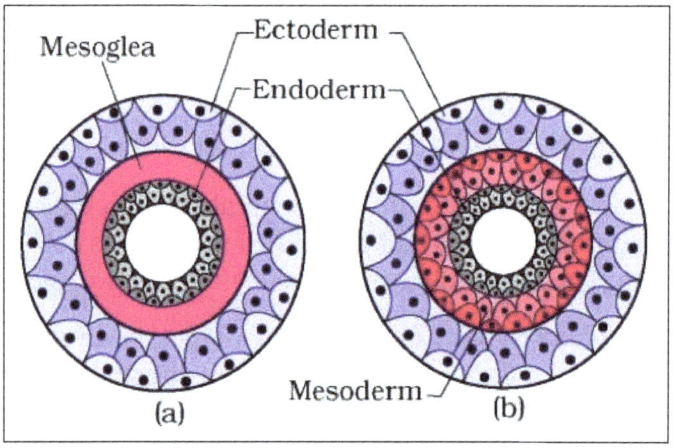

Showing germinal layers: (a) Diploblastic (b) Triploblastic.

Coelom

- Presence or absence of a cavity between the body wall and the gut wall is very important in classification.

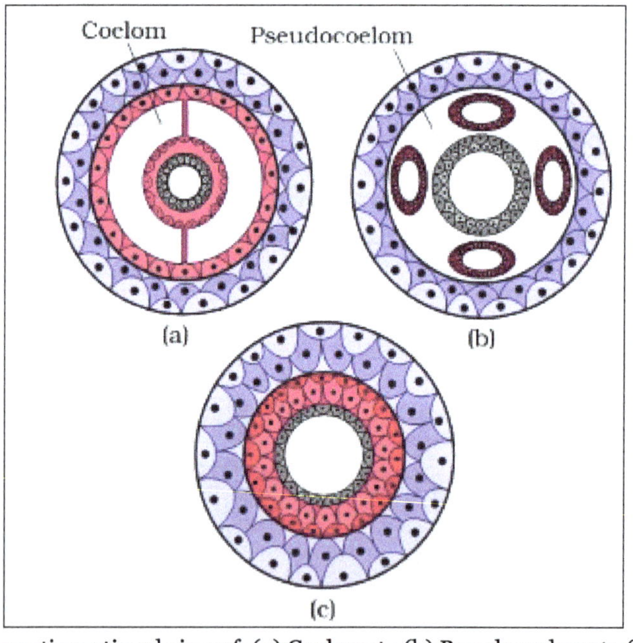

Figure: Diagrammatic sectional view of: (a) Coelomate (b) Pseudocoelomate (c) Acoelomate.

- The body cavity, which is lined by mesoderm, is called coelom.
- Animals possessing coelom are called coelomates, e.g., Annelids, Molluscs, Arthropods, Echinoderms, Hemichordates & Chordates.
- In some animals, the body cavity is not lined by mesoderm; instead, the mesoderm is present as scattered pouches in between the ectoderm and endoderm. Such a body cavity is called pseudocoelom and the animals possessing them are called pseudocoelomates, e.g., Aschelminthes.
- The animals in which the body cavity is absent are called acoelomates, e.g., Platyhelminthes.

Segmentation

- In some animals, the body is externally and internally divided into segments with a serial repetition of at least some organs.
- For example, in earthworm, the body shows this pattern called metameric segmentation and the phenomenon is known as metamerism.

Notochord

- Notochord is a mesodermally [the middle layer of cells or tissues of an embryo, or the parts derived from this (e.g. cartilage, muscles, and bone)] derived rod-like structure formed on the dorsal side [posterior] during embryonic development in some animals.
- Animals with notochord are called chordates and those animals which do not form this structure are called non-chordates, e.g., Porifera to Echinoderms.

Animal Phylum

Phylum is a taxon used in the scientific classification of life. "Phyla" is adopted from the Greek phylai, the clan-based voting groups in Greek city-states. Phyla represent the largest generally accepted groupings of animals and other living things with certain evolutionary traits, although the phyla themselves may sometimes be grouped into superphyla (e.g. Ecdysozoa with eight phyla, including arthropods and roundworms; and Deuterostomia with the echinoderms, chordates, hemichordates and arrow worms). Informally, phyla can be thought of as grouping animals based on general body plan, e.g., having bones vs. having an exoskeleton. This is morphological grouping. But despite the seemingly different external appearances of organisms, they are classified into phyla based on their internal organizations. For example, though seemingly divergent, spiders and crabs both belong to Arthropoda, whereas earthworms and tapeworms, similar in shape, are from Annelida and Platyhelminthes, respectively. Although the International Code of Botanical Nomenclature allows the use of the term "Phylum" in reference to plants, the term "Division" is almost always used by botanists.

The best known animal phyla are the Mollusca, Porifera, Cnidaria, Platyhelminthes, Nematoda, Annelida, Arthropoda, Echinodermata, and Chordata, the phylum humans belong to. Although there are approximately 35 phyla, these nine include the majority of the species. Many phyla are exclusively marine, and only one phylum is entirely absent from the world's oceans: the Onychophora or velvet worms. The most recently discovered phylum is Cycliophora found in 1993; only three phyla were discovered in the last century.

Phylum Porifera (Sponges)

Simple multicellular animals that spend their lives anchored to a rock or ocean bottom; they are mostly marine but some species live in fresh water; sponges have radial symmetry with a cylindrical, globose or irregular body containing an internal skeleton of minute spicules made of calcium carbonate, silica, or a fibrous collagen protein called spongin; the surface contains numerous pores connecting to canals and chambers lined by flagellated collar cells (choanocytes); sponges are filter-feeders, taking in microscopic plankton by miniature currents created by the choanocytes; sponges are divided into three classes: Class Calcarea (chalk sponges) with calcareous spicules; Class Hexactinellida (glass sponges) with siliceous spicules; and Class Demospongiae (horn sponges) with a skeleton of spongin or none; calcareous and siliceous sponge spicules often become components of tropical beach sand.

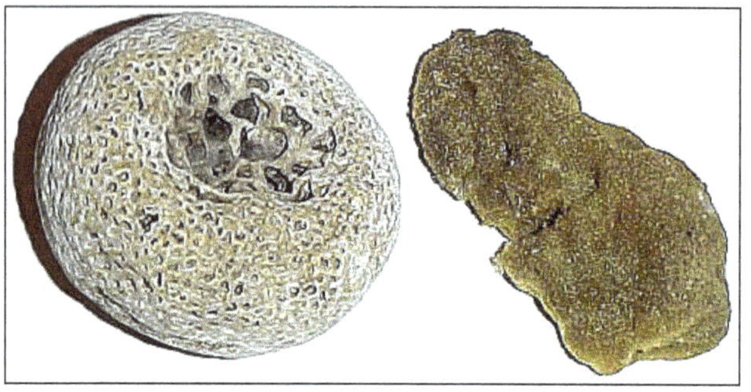

Two examples of marine sponges: The surface contains numerous pores connected to canals and chambers lined by flagellated cells called choanocytes. Sponges are filter-feeders, taking in microscopic plankton by miniature currents created by the choanocytes:

- Enterozoa: Animals with a digestive cavity or tract.
- Radiata: Radial symmetry, lacking central nervous system.

Phylum Coelenterata (Cnidaria) Corals and Jellyfish

Marine and freshwater animals with radial symmetry; this phylum has two distinct body forms: a solitary or colonial polyp and a bell-shaped, free swimming medusa; both polyps and medusae are often fringed with stinging tentacles; the tentacles bear rows of "stinging cells" or cnidoblasts, each containing a "stinging organelle" known as a nematocyst; some coelenterates (Obelia) have both polyp and medusa stages in their diplontic life cycle; coelenterates are divided into three classes: Class Hydrozoa (hydroids), including the freshwater Hydra, Obelia and colonial Physalia (Portuguese man-of-war) with stinging tentacles and a bladderlike air float; Class Scyphozoa (jellyfishes); and Class Anthozoa (sea anemones and corals); in true corals, the individual tentacle-bearing polyp lives inside a calcareous chamber that it secretes; extensive, contiguous colonies of coral polyps form massive calcareous reefs in shallow tropical waters; coral reefs support entire animal communities that rely on the photosynthetic activity of unicellular zooxanthellae that live within the coral polyps; no other ecosystem other than the tropical rain forest rivals coral reefs in terms of complexity and productivity.

Zoology textbooks often include a fourth class of cnidarians known as the box jellyfishes (class Cubozoa). These jellysish have a characteristic squarish medusa. The class Cubozoa includes the deadly Australian sea wasp (Chironex fleckeri). If deadliest venom is measured by how long it takes a person to die, then this species is certainly one of the world's deadliest. Depending on the extent of envenomation, the sea wasp can inflict excruciating stings that may cause death within five minutes. Ingenious Australian swimmers and snorkelers have devised a complete nylon body suit made from two pairs of panty hose, one put on the usual way; and the other put on over the hands, arms, and torso, with a slit for the head. Curiously, the deadly nematocysts of Chironex do not sting through panty hose, but watch out for runs.

The toxin of jellyfish nematocysts is a complex mixture of proteins, enzymes, polypeptides and tetramine. Pain and local histamine release of some jellyfish stings are attributed to 5-hydroxytryptamine, one of the ingredients in the stinging trichomes of nettles. The mechanism causing cardiac and respiratory failure of sea wasp venom is unclear. It may be due to blockage of nerve impulses at the synaptic junctions, similar to the action of curare.

A marine stony coral showing the numerous calcareous chambers that were once occupied by tentacle-bearing polyps. Colonies of stony corals are important reef builders in warm, tropical waters. Reef corals may form (1) fringing reefs extending out to 0.4 kilometers from shore; (2) barrier reefs separated by a lagoon of considerable width and depth from a shore; and (3) atolls or circular reefs that encircle a lagoon of water and not enclosing an island.

A pristine tide pool along the Oregon coast of North America. A. Purple sea urchin, B. Blood Starfish, C. Coralline red algae, D. Six-rayed starfish, and E. Sea anemone. The tentacles of the sea anemone contain zoochlorellae, symbiotic green cells of the algal division Chlorophyta.

Small, brightly colored fish called damselfish or anemone fish have evolved a mutualistic symbiotic association with large sea anemones. A slimy, mucus coating on the fish's body in some way inhibits the discharge of nematocysts along the sea anemone's tentacles. These little fish gain protection from predators by swimming down among the tentacles. The anemone benefits from its symbiont's movements, which dislodge silt and attract other fish into the lethal thicket of tentacles.

A sea anemone and its symbiotic anemone fish.

Phylum Ctenophora (Comb Jellies)

Solitary marine animals with transparent biradially symmetrical bodies; externally there are eight plates of fused cilia that resemble long combs; the rows of ciliated comb plates are used for locomotion; most of the body is composed of a jellylike material called mesoglea; the digestive (gastrovascular) cavity has branched canals; unlike the coelenterates, comb jellies do not have nematocysts; an Atlantic coastal species (Mnemiopsis) exhibits bioluminescence, the emission of light by on organism or population of organisms. This phenomenon involves the oxidation of luciferin in the presence of ATP and the enzyme luciferase. Examples of bioluminesence include dinoflagellates causing "red tide," lightning "bugs" (beetles), glow worms (beetle larvae) and the deep sea angler fish.

Three comb jellies (ctenophores). Comb jellies resemble tiny hot air ballons (the size of a walnut or smaller) with eight rows of fused cilia (comb plates) extending down the sides. They propel themselves mouth first by the eight rows of comb plates. Ctenophores superficially resemble miniature medusae (phylum Cnidaria); however, most medusae arise asexually from a polyp generation and ctenophores have no polyp stage in their life cycle. Tentacles extending from the mouth contain "glue cells" or colloblasts containing spiral threads which snare small fish and crustaceans with a gluelike material. With the exception of one species, ctenophores do not have the stinging organelles (nematocysts) of jellyfish and sea anemones. During the day, ctenophores flash prismatically as their ciliary plates refract light; at night they are often bioluminescent, glowing like little lamps:

- Bilateria: Bilateral symmetry, with central nervous system. Phylum Echinodermata with radial symmetry.
- Acoelomata: No body cavity (coelom).

Phylum Platyhelminthes (Flatworms)

Body flattened, leaf or ribbonlike, bilaterally symmetrical; digestive tract branched and without an anus, or absent in parasitic forms; this phylum is divided into three classes: Class Turbellaria (free-living flatworms); Class Trematoda (flukes); and Class Cestoda (tapeworms).

Life Cycle of a Human Tapeworm

A man eats a poorly cooked piece of beef containing the encysted bladderworm (tapeworm cyst) in the skeletal muscle tissue. The young tapeworm attaches to the man's intestinal wall by its smaller, anterior end called the scolex. The scolex anchors the the tapeworm to the host's intestinal wall by four suckers. Some early diet pills actually contained a tapeworm cyst enclosed in a gelatin capsule, before they were banned by the Food & Drug Administration. You could take the pill and eat as much as you want without gaining weight! Of course, you were actually feeding a large tapeworm which could be very dangerous. The tapeworm grows by budding as the scolex produces flattened segments called proglottids. A human tapeworm may contain hundreds of proglottids and be several feet in length. Tapeworms compete with their host for nutrients. Large tapeworms may obstruct food passage, and if their scolex penetrates the intestinal lining, bacterial infection may result.

Tapeworms are hermaphroditic and each mature proglottid contains a complete male and female reproductive system. Self-fertilization may occur, or cross-fertilization between proglottids of different worms may also occur. A single ripe proglottid may contain more than 100,000 eggs. The annual output of one worm is about 500 million eggs and the worm may continue this production for 20 years or more. Proglottids filled with embryonated eggs (containing embryos) break off and are discharged with human feces.

If human sewage is discharged on grazing land, the embryo-bearing, tapeworm eggs may be consumed by cattle while grazing. Within the cattle, the larva hatches into a six-hooked oncosphere stage which penetrates the gut wall and eventually ends up in striated muscle tissue where it encysts. Within the cyst, the larva develops into a bladder-worm, a miniature tapeworm complete with scolex and attachment structures. If striated muscle tissue (poorly cooked beef) is eaten by another person, an encysted bladder-worm may take up residency in the new person's intestine. There are literally hundreds of different species of tapeworms, and they infect cattle, sheep, dogs, cats, pigs, fish and many other animals.

Much more serious than infection with an encysted bladder-worm is the ingestion of tapeworm eggs directly. If the eggs are swallowed, the oncosphere larvae migrate through the body, not only skeletal muscle, but also into the heart, brain and other organs. Larvae of the dog tapeworm (Dipylidium caninum) live inside the bodies of fleas and lice. If eggs from this dangerous tapeworm are ingested, the worm may invade vital organs such as the brain. The habit of kissing the lips and tongue of a dog is somewhat questionable, especially considering their habit of biting fleas and licking themselves and other dogs in anal-genital regions. Another story to "gross you out" involves an affectionate cat that liked to rub against my arms while petting her. One day I noticed a tiny flat object on my elbow that just happened to be a fresh proglottid from a tapeworm inside her intestine.

Human tapeworms may be several feet long. In fact, one reported case at a medical clinic in San Diego County is astonishing. A man noticed a long white ribbon in his feces and thought he had swallowed his pajama draw strings during the night. The white ribbon turned out to be a tapeworm.

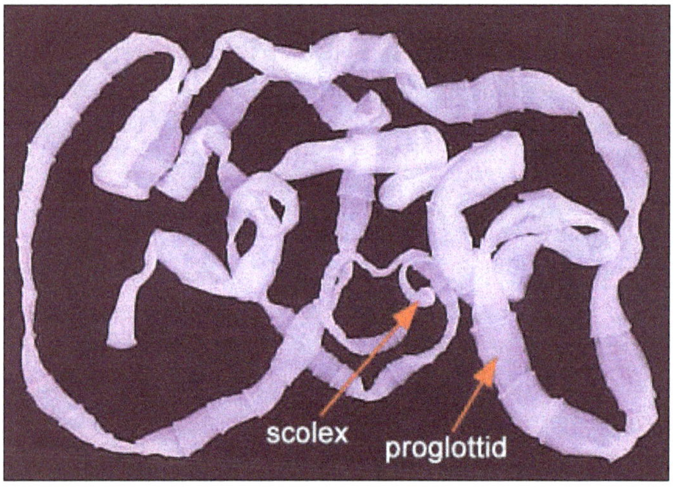

Is this a (1) wet pajama draw string, (2) a long noodle, or (3) a human tapeworm? The answer is (3), a 20 inch (50 cm) human tapeworm. The small head or scolex from which the segments (progottids) arise is clearly visible in the photograph. The presence of a scolex is good evidence that the entire worm has been discharged from the host's intestine. Each proglottid contains a complete male and female reproductive system. In fact, one proglottid may contain literally thousands of eggs.

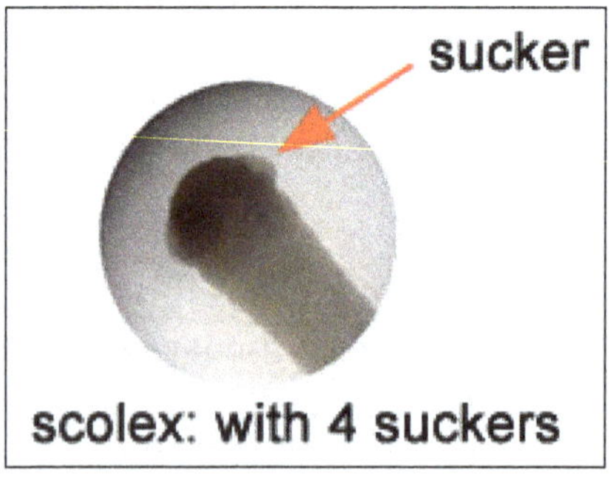

The human tapeworm shown in the above photograph showing the anterior end or scolex. The scolex bears four circular suckers which firmly anchor the tapeworm to the host's intestinal wall. The scolex produces proglottids by budding, which gradually enlarge as more segments are formed.

Phylum Nemertea (Ribbon Worms)

Free-living marine worms with a distinctive eversible proboscis consisting of a long, hollow tube, the unsegmented body is covered with cilia; like flatworms, arrow worms are acoelomates (without a coelom), but they do have a complete digestive system. Pseudocoelomata: Coelom incomplete.

Phylum Rotifera (Rotifers)

Rotifers are common microscopic animals living in fresh water, particularly among plants and debris; the rotating cilia at the anterior end resemble the head of an electric shaver.

Phylum Gastrotricha (Gastrotrichs)

Microscopic aquatic animals similar to the rotifers, but with a more flattened body; they are planktonic organisms living among algae and debris.

Phylum Nematomorpha (Horsehair Worms)

The adults are very long, thin, hairlike worms that are free-living in water; the immature (larval) stage is usually parasitic on insects; people once thought that dark hairs from a horses tail at a watering hole miraculously came to life; these so-called "hairs" were actually horsehair worms.

Phylum Nematoda (Nematodes)

A large phylum of worms, often called round worms or thread worms; free living in soil or water, with many parasitic forms on plants and animals, including hookworms (Necator & Ancylostoma), pinworms (Enterobius), human ascaris Ascaris lumbricoides and trichinosis worm Trichinella; hookworm larvae burrow into the skin between the toes and eventualy migrate into the intestines; the disease trichinosis is caused by eating undercooked meat (especially pork) containing microscopic cysts of larval Trichinella spiralis; if ingested the larval cysts can eventually migrate into skeletal muscle tissue; elephantiasis is caused by minute filarial worms of the genus Wuchereria that invade lymphatic vessels; the filarial worms obstruct the flow of lyphatic fluids causing body extremities to swell to gargantuan proportions; microscopic filarial larvae causing elephantiasis are spread by blood-sucking female mosquitoes; other dreaded parasitic worms in the phylum include the Guinea worm (Dracunculus medinensis) and the giant kidney worm (Dioctophyma renale).

Phylum Acanthocephala (Spiny-Headed Worms)

Small round worms that are intestinal parasites of vertebrates, usually with a crustacean or insect as an alternate host in young stages of its life cycle; a proboscis at the anterior end is covered with hooks for attachment to the wall of the intestines. Eucoelomata: Body with a true coelom.

Phylum Bryozoa (Bryozoans)

Bryozoans form coral-like colonies in the ocean and fresh water; they are often called "moss animals" because of their delicate branched colonies; each animal has a fringe of tentacles around its mouth, and occupies a separate chamber similar to the individual chambers of coral polyps; freshwater bryozoans produce disk-shaped, chitinous reproductive bodies called statoblasts which are sometimes very abundant is wolffia samples. Rotostomia: Blastopore forms mouth, schizocoelom present.

Phylum Tardigrada (Tardigrades)

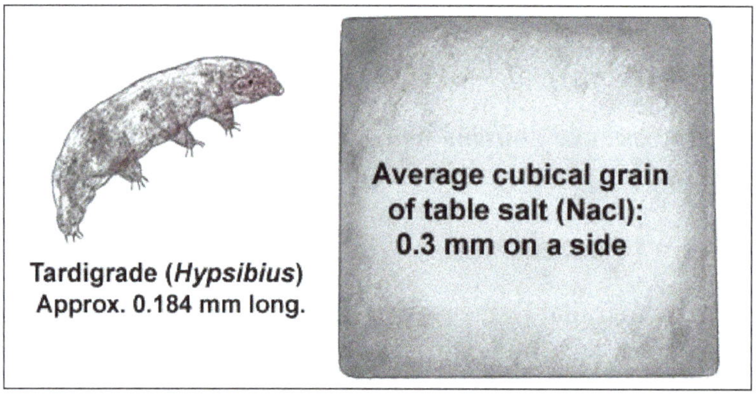

Tardigrades belong to a remarkable phylum of minute multicellular animals. They are adapted to extreme conditions, some of which are more severe than any earth environment. Does their origin defy natural selection?

Tardigrades are microscopic, water-dwelling, segmented animals with eight legs. Depending on the species, they range in size from 0.1 to 1.5 mm long. They are called "water bears," in reference to their stumpy legs tipped with claws, and especially their lumbering gait. More than 1,000 species have been described. They occur throughout the world in some of the most inhospitable places, from high peaks in the Himalayas (above 20,000 ft) to boiling water near heat vents on the ocean floor, and from the Arctic tundra to vast ice fields of Antarctica. During severe environmental conditions that would kill most creatures on earth, tardigrades roll up into little dehydrated balls called "tuns" where they survive for extended periods of time. This dormancy phenomenon is called cryptobiosis (or anabiosis). Their ability to survive in this desiccated state for long periods is largely dependent on high levels of the non-reducing sugar trehalose that protects their tissues and DNA. Tardigrade physiology has led to the discovery of "dry vaccines" that don't require refrigeration and thus can be delivered and stored at room temperature. Some species can survive temperatures near absolute zero (-273 °C) where liquids and gasses freeze solid; some can survive temperatures up to 151 °C (304 °F); some can live without water for 10 years; some can survive 1,000 times more gamma radiation than other animals (tardigrades can withstand 570,000 rads of X-ray radiation while 500 rads would kill a human); some can withstand pressures up to 6,000 atmospheres (more than the deepest ocean trench), and some can actually live for a while in a vacuum. In fact, tardigrades were taken into earth orbit on the Russian robotic spacecraft FOTON-M3 and survived 10 days in space. Upon returning to earth and hydrated, many actually laid eggs that hatched normally.

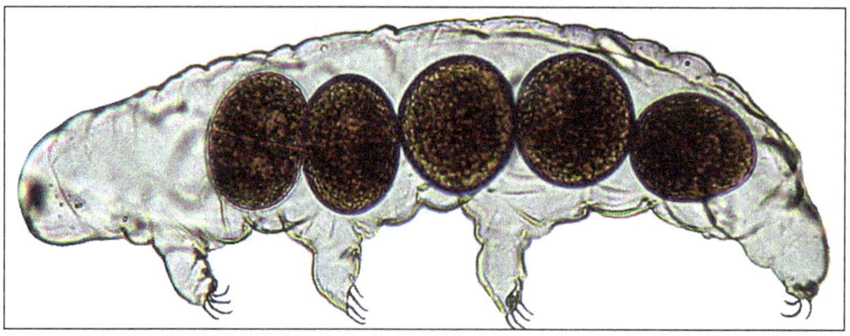

Lateral (side) view of the exoskeleton of an aquatic tardigrade (Hypsibius sp.) containing 5 eggs. There are 4 pairs of stout, stumpy legs, each tipped with several slender claws. The name "water bear" refers to its deliberate "pawing" sort of locomotion. Thomas Huxley, English naturalist and good friend of Charles Darwin, gave tardigrades this name in 1869.

Advocates of intelligent design believe that tardigrades defy evolution by natural selection because they are adapted to extreme conditions, some of which are more severe than any earth environment. Creationists maintain that natural selection can only select characteristics necessary for immediate survival. Consequently, evolution cannot be expected to "over-equip" species for a host of environments that they have never faced. Advocates of the panspermia hypothesis believe that tardigrades had an extraterrestrial origin, although DNA sequencing data indicate that the phylum Tardigrada (tardigrades) is a sister clade with the phylum Arthropoda (arthropods), and both phyla (Tardigrada + Arthropoda) form a sister clade with the phylum Onychophora (velvet worms).

A strong case to support the evolution of "over-equipped" species is "gene duplication" in diploid organisms that greatly increases genetic variability. One of the truly remarkable examples of gene duplication is the antibody mediated immune system of animals. Gene duplication is a plausible explanation for how organisms can produce antibodies against different antigens, even synthetic antigen proteins that animals have never been exposed to. Using this model, animals would not need separate genes for every antigen they will ever encounter. This mechanism goes way beyond the simple evolution of adaptations based on Darwin's theory of evolution by means of natural selection.

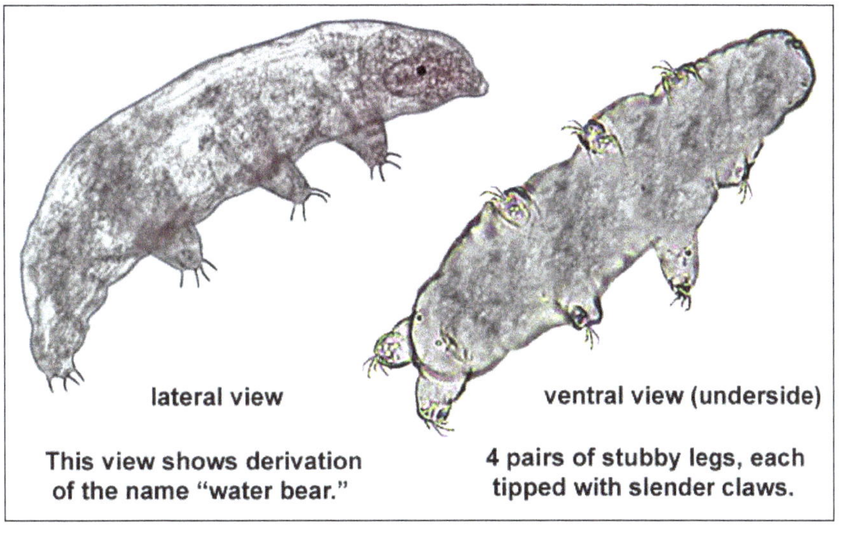

An aquatic tardigrade of the genus Hypsibius. Its length is approximately 184 micrometers (microns), about the same length as the hair follicle mite Demodex brevis. It is much smaller than a grain of common table salt (NaCl). The image was enhanced with Photoshop to bring out detail of the claws.

Phylum Brachiopoda (Brachiopods)

Primitive marine bivalve animals attached to rocks by a fleshy stalk; brachiopod shells are abundant in the fossil record, dating back to the Cambrian Period (500 million years ago). There are approximately 280 species of living brachiopods, only a fraction of the 30,000 described fossil species that flourished during the Paleozoic and Mesozoic eras. My college invertebrate paleontology textbook contains a lengthy chapter dedicated to the Brachiopoda, with literally hundreds of illustrations of extinct brachiopods.

Although they superficially resemble bivalve mollusks, they are really quite different. In brachiopods, the two valves enclose the body dorsally and ventrally instead of laterally as in mollusks. The bivalve shells may be hinged (articulate) or without hinges (inarticulate). In addition, the ventral valve is usually larger than the dorsal, and is usually attached to the substrate directly or by means of a cordlike stalk.

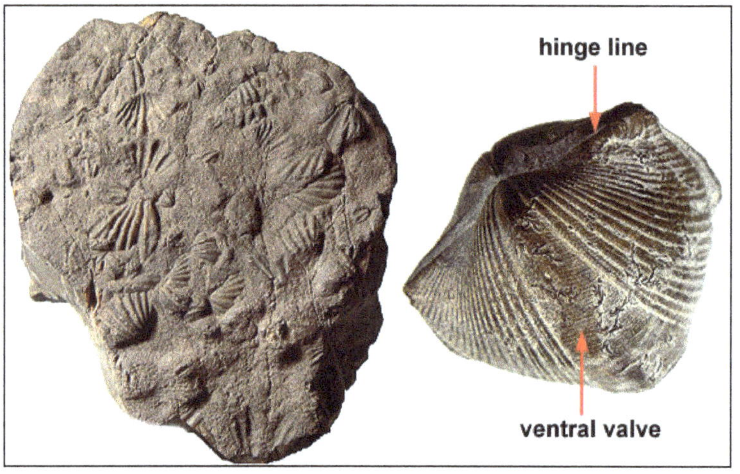

Left: Mesozoic rock from Arizona showing numerous brachiopods, evidence of a shallow sea along a continental shelf. Right: An articulate (hinged) brachiopod.

Phylum Mollusca (Mollusks)

Body soft with bilateral symmetry, often covered by a mantle that secretes a calcareous shell; usually with an anterior head and a ventral muscular foot for locomotion; divided into five classes: Class Amphineura (chitons); Class Scaphopoda (tooth shells); Class Gastropoda (univalve mollusks including snails & limpets); Class Pelecypoda (bivalve mollusks); and Class Cephalopoda (octopus & squid).

Gastropods shells (cowries and cone shells) from the south Pacific island of Moorea and Tetiaroa Atoll in French Polynesia. The lower right shell is from a land snail.

The common garden snail (Helix aspersa). This snail was introduced into California in the 1850s by European immigrants for use as food. It turned out to be a poor substitute for the edible snail (escargot) served in French restaurants (Helix pomatia). It has since become well-established and a major nuisance in California gardens. Homeowners spend more money on eradicating this creature than all other invertebrate pests combined. The head has two pairs of retractile tentacles. The longer, upper tentacles each have an eye at the tip. The shorter, lower tentacles bear chemosensory organs (equivalent to taste and smell). The ventral side contains a mouth with a rasping tongue (radula) for feeding. This snail is hermaphroditic (bisexual) with both male and female sex organs. Although capable of self-fertilization, the normal method is cross fertilization between two mating partners. The snail moves by the gliding action of its large muscular foot. Mucus secreted from slime glands on the foot provides a lubricant to facilitate movement, and leaves a silvery slime trail. Snail mucus is used in "snail cream" cosmetics. The cream is used to treat damaged skin and as a skin moisturizer.

Phylum Annelida (Segmented Worms)

Body elongate and composed of many segments, each segment with fine bristlelike setae for locomotion; divided into three classes: Class Polychaeta (clamworms); Class Oligochaeta (earthworms); and Class Hirudinea (leeches).

The formidable bloodworm (Glycera), a predaceous intertidal worm (phylum Annelida) with four fanglike jaws at the anterior end (red arrow). The hollow fangs inject poison from venom glands into its prey. The elongate body is composed of many segments, each segment with fine bristle-like setae for locomotion. This interesting worm lives in tubular gallaries constructed in muddy bottoms. Lying in wait within its tube, the worm can detect the surface movements of prey by changes in the water pressure. It slowly moves to the opening of its burrow and then seizes the prey, such as small crustaceans and other invertebrates, with its proboscis. The proboscis can be extended to about one fifth the length of its body. When retracted, the proboscis occupies approximately the first 20 body segments. During the capture of prey, the proboscis is everted with explosive force, and the four venom fangs emerge at the tip.

Phylum Sipunculoidea (Peanut Worms)

Body slender, gourd-shaped; anterior end retractile, with short hollow tentacles around the mouth.

Peanut worms (phylum Sipunculoidea) are named after their shape when when fully contracted, such as when they are handled. These interesting worms live in burrows under rocks of the intertidal zone. They are sometimes very abundant in tunnels they bore in coralline rocks, reportedly up to 700 per square meter in some Hawaiian reefs. The slender anterior end has a mouth surrounded by hollow tentacles.

Phylum Arthropoda

Body composed of head, thorax and abdomen with three or more pairs of jointed legs; chitinous exoskeleton covering all body parts, molted at intervals; divided into seven major classes: Class Onychophora (walking worms); Class Crustacea (shrimp, crabs & barnacles); Class Insecta (insects); Class Chilopoda (centipedes); Class Diplopoda (millipedes); Class Arachnida (spiders & ticks); and Class Merostomata (horseshoe crabs).

Ostracods Belong to the Arthropod Class Crustacea Order Ostracoda

These microscopic objects are not eggs. They are the dead, chitinous, bivalve carapaces of minute crustaceans called mussel or seed shrimps. When they were alive, a tiny shrimplike animal occupied each white case. They belong to the order Ostracoda which is considered a class in some textbooks. Another order of minute crustaceans (Conchostraca) are called clam shrimp; however, the bivalve carapace of an ostracod is more strikingly similar to a clam shell. Photo was taken along the shore of a desiccated vernal pool in the Santa Rosa Plateau of Riverside County, California.

Brine Shrimp Belong to the Arthropod Class Crustacea Order Anostraca

A thriving population of brine shrimp (Artemia monica) in Mono Lake, California. These minute crustaceans belong to the order Anostraca, along with the endangered fairy shrimp (Branchinecta). During the summer months, the shoreline of Mono Lake is colored pinkish-red by massive colonies of brine shrimp. The Phalarope (right) is swimming in circles as it feeds on the brine shrimp.

A common southern California centipede (Scolopendra polymorpha).

The name centipede is derived from the Latin words "centi" (100) and "pede" (foot). Centipedes belong to the class Chilopoda in the phylum Arthropoda. Adult centipedes in North America may have fewer or more than 100 legs, one pair per body segment. Centipedes are the only animals with legs modified into fanglike "poison jaws" (toxicognaths) that inject poison for subduing and killing prey. Modified forelegs, called prehensors, are located under the head. Glands inside the prehensors release venom into ducts that lead into the fangs. The last pair of legs is longer than the others. Some species use these to subdue prey or as defensive pincers. The posterior body segment with its unique pair of legs superficially resembles the head. This “pseudohead” found in some species may serve to confuse potential predators by misdirecting their attacks to a less vital part of their body, thus leaving the head free to bite the attacker. A common question that students ask is “can centipedes bite?” we can verify that they indeed have powerful fangs and can inflict a painful bite, particularly larger species shown in the following two images.

Magnified view of the underside of a centipede (Scolopendra polymorpha) showing a pair of fanglike “poison jaws” or toxicognaths (red arrow). These are actually modified forelegs that inject venom.

A large southern California millipede (probably Hiltonius pulchrus), a member of the class Diplopoda in the phylum Arthropoda. Diplopoda means “double-footed,” and refers to two pairs of legs on each body segment. The name millipede is derived from the Latin words “milli” (1000) and “pede” (foot). Actually, adult centipedes in North America may have up to 750 legs, typically two pairs per body segment. Unlike centipedes, most millipedes are harmless detritis feeders (detritivores) and feed on rotting vegetation.

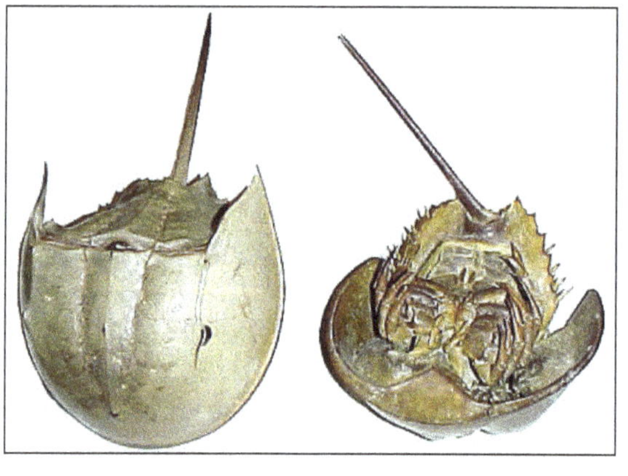

The horseshoe crab (Limulus polyphemus) lives in shallow water along the Atlantic coast and the Gulf of Mexico. It belongs to the class Merostomata in the subphylum Chelicerata. The horseshoe crab is a living representative of the subclass Xiphosura, and line of aquatic chelicerates that dates back to ancient Cambrian seas 500 million years ago. The subphylum chelicerata is characterized by arthropods with a cephalothorax, no antennae and jaws called chelicerae. In addition to the horseshoe crab, this subphylum includes the large class Arachnida (spiders, scorpions, ticks and mites). Another large subphylum, the Mandibulata, is characterized by a head, thorax and abdomen, paired antennae and jaws called mandibles. It includes the class Crustacea (lobsters, crabs and barnacles), class Insecta (insects), class Chilopoda (centipedes) and class Diplopoda (millipedes).

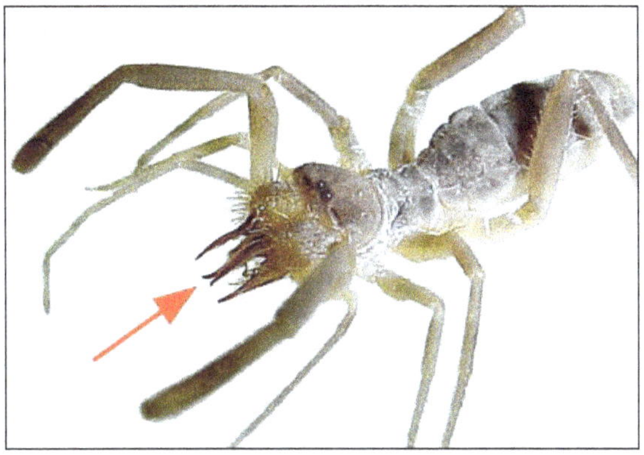

A sun spider or solpugid (order Solpugida), an interesting member of the subphylum Chelicerata in the class Arachnida. Unlike true spiders (order Araneae), the solpugid has two pairs of formidable, scissorlike jaws called chelicerae (red arrow). The jaws are arranged vertically rather than the single pair of horizontal jaws in most spiders. The elongate pair of anterior appendages are pedipalps, each tipped with a suckers that are used for capturing prey and climbing smooth surfaces. Solpugids are extremely voracious carnivores, even attacking and killing other large, webless spiders, such as wolf spiders. With their twin pairs of huge, powerful jaws, solpugids literally crush and tear their prey to shreds.

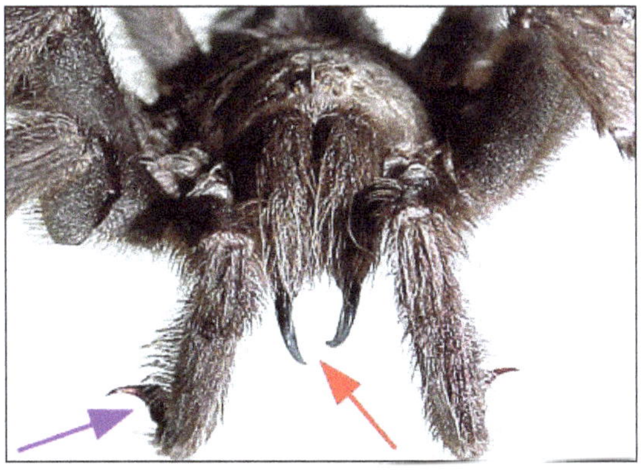

Close-up view of the anterior end of a tarantula (Aphonopelma) showing two sharp, fangs (red arrow) extending from the jaws or chelicerae. This spider is a male because its pedipalps are tipped with a sharp-pointed, bulblike copulatory organ (purple arrow). During mating the male inserts sperm into an opening on the underside of the female using the pointed copulatory organs on his pedipalps. Several closely-spaced, orange eyes can be seen on the front of the head (cephalothorax). The tarantula kills its prey by crushing and injecting them with poison through its hollow fangs. Although it is a large spider, the tarantula is hunted and killed by a large black wasp with bright red wings known as the tarantula hawk (Peplis). Deuterostomia: Blastopore forms anus, enterocoelom present.

Phylum Chaetognatha (Arrow Worms)

Small, transparent marine worms with bristles or hooks around the mouth; they have paired fins on the main body trunk and a terminal tail fin for locomotion; although very small, they have well-developed digestive and sensory systems, and a true coelom.

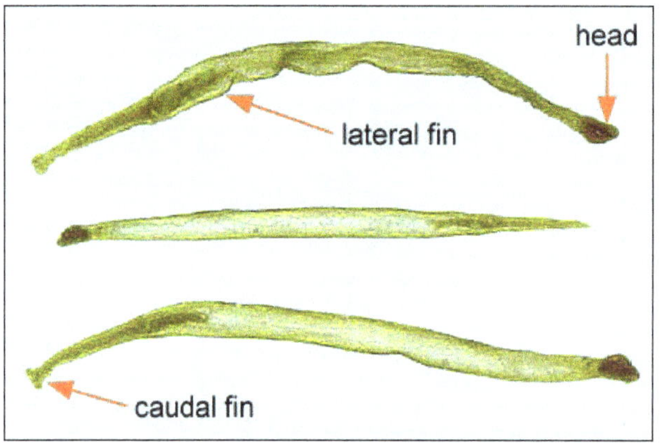

Close-up view of several arrow worms (phylum Chaetognatha). These planktonic marine animals are only about 5-10 mm in length, about the size of duckweeds of the genus Wolffiella. The have a torpedo-shaped body with lateral and caudal (tail) fins that superficially resemble the feathers of an arrow. The head is covered by a rounded hood that is retracted when the arrow worm feeds. During feeding, the arrow worm captures small plantonic organisms, including copepods and other chaetognaths, with minute spines or bristles around the mouth.

Phylum Echinodermata (Echinoderms)

Body radially symmetrical, usually 5-parted around an oral-aboral axis; body wall with calcareous plates, often with external spines; coelom includes water vascular system and external tube feet for locomotion; divided into five classes: Class Crinoidea (sea lilies); Class Asteroidea (starfish); Class Ophiuroidea (brittle stars); Class Echinoidea (sea urchins & sand dollars) and Class Holothuroidea (sea cucumbers).

Two well-preserved, fossilized echinoderms. Left: A 40 million-year-old (Eocene Epoch) sea biscuit (Eupatagus ocalanus) from Ocala limestone (Marion County, Florida). Right: A 7 million-year-old

(Pliocene Epoch) sand dollar (Millitella californicus) from Baja California, Mexico. Both of these organisms belong to the class Echinoidea along with sea urchins.

Phylum Hemichordata (Acorn Worms)

A small group of marine invertebrates that were once thought to have a short notochord, the supportive axial rod that characterizes the phylum Chordata; hemichordates do possess gill slits or pharygeal clefts that are also found in chordates.

Phylum Chordata (Chordates)

Having at some stage of development an axial rod-like notocord for support of the body; chordates also have a single dorsal tubular nerve cord, paired gill slits between the pharynx and exterior, and a tail behind the anus; subdivided into three major subphyla: Subphylum Urochordata (tunicates); Subphylum Cephalochordata (lancelets); and Subphylum Vertebrata (vertebrates).

Vertebrates: Subphylum Vertebrata: With a cranium (skull) and spinal column composed of segmented vetebrae; skeleton composed of cartilage or true calcareous bones; divided into eight major classes: Class Agnatha (jawless fishes); Class Placoderms (armored fishes); Class Chondrichthyes (cartilaginous fishes including sharks and rays); Class Osteichthyes (bony fishes); Class Amphibia (amphibians); Class Reptilia (reptiles); Class Aves (birds); and Class Mammalia (mammals).

Garibaldi (Hypsypops rubicundus), a bright orange fish of the shallow waters off the coast of California. Like California's state flower (golden poppy) and state freshwater fish (golden trout), the garibaldi has been selected to be the state's official marine fish.

Animal Class

An animal class is made up of animals that are all alike in important ways. Scientists have grouped animals into classes to make it easier to study them. There are many different animal classes and every animal in the world belongs to one of them. The five most well-known classes of vertebrates (animals with backbones) are mammals, birds, fish, reptiles, amphibians. They are all part of the phylum chordata - We remember "chordata" by thinking of spinal cord. There are also a lot of animals without backbones. These are called invertebrates and are part of the phylum arthropoda (arthropods). Two of the most commonly known classes in this phylum are arachnids (spiders) and insects.

Vertebrates

Animals that have a backbone are called vertebrates. The backbone is a column of bones that supports the body, and also protects the spinal cord. The back is made up of many separate bones, called vertebrae. The backbone is part of a vertebrate's supportive internal skeleton. Most vertebrates have two pairs of limbs. In fish these take the form of fins, but in other vertebrates they are legs, arms, flippers, or wings. Some vertebrates, such as snakes, have no external limbs. Animals with backbones include mammals, birds, reptiles, amphibians, and fish.

Fishes

Whale shark: A whale shark (*Rhincodon typus*) and a snorkeler off the coast of Australia.

The first fishes are thought to have emerged some 518 million years ago during the Cambrian Period of Earth's history. Today, there more than 30,000 species of fishes found in the fresh and salt waters of the world. Living species range from the primitive, jawless lampreys and hagfishes through the cartilaginous sharks, skates, and rays to the abundant and diverse bony fishes.

Fishes range in adult length from less than 10 mm (0.4 inch) to more than 20 meters (60 feet) and in weight from about 1.5 grams (less than 0.06 ounce) to many thousands of kilograms. Some live in shallow thermal springs at temperatures slightly above 42 °C (100 °F), others in cold Arctic seas a few degrees below 0 °C (32 °F) or in cold deep waters more than 4,000 meters (13,100 feet) beneath the ocean surface.

Fish reproduction methods vary, but most fishes lay a large number of small eggs that are fertilized and scattered outside of the body. The eggs of pelagic (open ocean) fishes usually remain suspended in the open water, while many shore and freshwater fishes lay eggs on the bottom or among plants. The mortality of the young and especially of the eggs is very high, and often only a few individuals grow to maturity out of hundreds, thousands, and in some cases millions of eggs laid.

Amphibians

Red salamander (*Pseudotriton ruber*) Red salamanders (*Pseudotriton ruber*) are found in the eastern United States.

Amphibians evolved from fully aquatic tetrapods—(which were essentially "limbed fish") who descended from lobe-finned fish—sometime between the Early Devonian Period (which began 419 million years ago) to the Early Pennsylvanian Subperiod (which began 323 million years ago). The name amphibian, derived from the Greek *amphibious* meaning "living a double life," reflects this dual life strategy—though some species are permanent land dwellers, while other species have a completely aquatic mode of existence.

There are three living groups of amphibians (caecilians, salamanders, and anurans [frogs and toads]) that, collectively, make up more than 7,300 amphibian species. One similar tendency among amphibians has been the evolution of direct development, in which the aquatic egg and free-swimming larval stages are eliminated. Development occurs fully within the egg capsule, and juveniles hatch as miniatures of the adult body form. Most species of lungless salamanders (family Plethodontidae), the largest salamander family, some caecilians, and many species of anurans have direct development. In addition, numerous caecilians and a few species of anurans and salamanders give birth to live young.

Frogs and toads display a wide variety of life histories. Some deposit eggs on vegetation above streams or ponds; upon hatching, the tadpoles drop into the water where they continue to develop throughout their larval stage. Some species create foam nests for their eggs in aquatic (watery), terrestrial (land-based), or arboreal (tree-based) habitats; after hatching, tadpoles usually develop in water. Other species deposit their eggs on land and transport them to water, while marsupial frogs are so called because they carry their eggs in a pouch on their backs. A few species lack a pouch and the tadpoles are exposed on the back; in some species, the female deposits her tadpoles in a pond as soon as they emerge from eggs.

Reptiles

Reptiles are air-breathing vertebrates. They have internal fertilization, amniotic development (in which the embryo develops within a set of protective extra-embryonic membranes—the amnion, chorion, and allantois), and epidermal scales covering part or all of their body. The major groups of living reptiles—the turtles, tuataras, lizards and snakes, and crocodiles account for over 8,700 species.

Male flying lizard.

Reptiles evolved from amphibians during the first part of the Pennsylvanian sub-period (323 million to 299 million years ago) and retained many amphibian structural characteristics. While most reptiles feed on other organisms, a few are herbivorous (e.g., tortoises). As cold-blooded animals, reptiles tend to be limited to temperate and tropical areas, but, where they occur, they are relatively common; however, they are not as large or conspicuous as birds and mammals. Most reptiles are terrestrial, but a few are aquatic. They move about by creeping or swimming in a fashion similar to amphibians. Some reptiles, however, can lift the body from the ground and run rapidly either in a quadrupedal or bipedal fashion. Reptiles lay relatively large, shelled eggs. In a few instances, the eggs and young are cared for by the female; in others, the young are born alive.

Birds

European white pelican: (*Pelecanus onocrotalus*) in flight.

Birds make up any of the 9,600 living species unique in having feathers, the major characteristic that distinguishes them from all other animals. They are warm-blooded vertebrates more related to reptiles than to mammals. They have a four-chambered heart (as do mammals), forelimbs modified into wings (a trait shared with bats), a hard-shelled egg, and keen vision. Their sense of smell is not highly developed, and their auditory range is limited.

Although most are capable of flight, others are sedentary, and some are flightless. In a manner similar to their relatively close relatives the reptiles, birds lay shelled eggs. The young are usually cared for in a nest until they are capable of flight and self-feeding, but some birds hatch in a well-developed state that allows them to begin feeding immediately or even take flight. (Nesting activities similar to those of some birds are seen in the crocodilians).

The origin of birds, feathers, and avian flight has long been hotly debated; the evolution of birds from reptilian ancestors is universally accepted, however. The diversity of theropod dinosaurs (a diverse group of carnivorous "lizard-hipped" dinosaurs), some with feathers, has greatly expanded our perspective of the evolution and early diversification of birds. While it is known that the critical period in avian evolution and flight took place during the Early Cretaceous (145.5 million to 99.6 million years ago), there is evidence that feathers on theropods emerged much earlier, possibly during the Triassic and Jurassic Periods (some 252 million to 145 million years ago).

Mammals

Okapi (*Okapia johnstoni*).

There are approximately 5,000 species of mammals living today. Mammals differ from other vertebrate animals in that their young are nourished with milk from special mammary glands of the mother. Mammals are distinguished by several other unique features. Hair is a typical mammalian feature, although in many whales it has disappeared except in the fetal stage. The mammalian lower jaw is hinged directly to the skull, instead of through a separate bone (the quadrate) as in all other vertebrates. A chain of three tiny bones transmits sound waves across the middle ear. A muscular diaphragm separates the heart and the lungs from the abdominal cavity. Mature red blood cells (erythrocytes) in all mammals lack a nucleus; all other vertebrates have nucleated red blood cells. The oldest known animals classified as mammals evolved near the boundary of the Triassic and Jurassic Periods, some 200 million years ago.

This group of vertebrates ranges in size from tiny shrews or small bats weighing only a few grams to the largest known animals, the whales. Most mammals are terrestrial, feeding on both animal and vegetable matter, but a few are partially aquatic or entirely so, as in the case of the whales or porpoises. Mammals move about in a great variety of ways: burrowing, bipedal or tetrapedal (four-legged) running, flying, or swimming. Reproduction usually involves the young developing inside the uterus, where nutritive materials are made available through an allantoic placenta or, in a few cases, a yolk sac. In placental mammals, young have a longer developmental period within the uterus. In marsupials, the relatively undeveloped young are carried in a pouch, where they attach themselves to their mother's nipple until they become fully developed. Monotreme mammals (that is, the platypus and echidna) differ from other mammals in that they lay eggs which hatch.

Invertebrates

Invertebrates are animals that don't have a backbone. The vertebral column is another name for the backbone. Over 90% of all species on Earth are invertebrates, and invertebrate species have been found in the fossil record as far back as 600 million years ago. Molecular biology studies suggest that all invertebrates evolved from a single invertebrate group.

Characteristics of Invertebrates

In addition to not having a backbone, invertebrates have soft bodies because they don't have an internal skeleton (endoskeleton) for support. Instead, many have structures on the outside (exoskeleton) that provides support and protection. In addition, invertebrates are cold-blooded, meaning they can't regulate their body temperature, so it changes depending on the environment.

Invertebrates are incredibly diverse. They live in fresh water, salt water, on land and as parasites in other animals. There are invertebrates that are carnivorous (meat eaters), herbivores (plant eaters) and omnivores (meat and plant eaters). There are even some invertebrate species that grow bacteria and cells inside their bodies that make their food. Some invertebrates stay in one spot, while others fly, swim, float, crawl and burrow.

Types of Invertebrates

Eighty-five percent of invertebrates – some 923,000 species – are arthropods. Mollusks have approximately 100,000 distinct species. Some of the most common types of invertebrates are:

- Protozoans: Single-celled organisms such as amoebas and paramecia.
- Annelids: Earthworms, leeches.
- Echinoderms: Starfish, sea urchins, sea cucumbers.
- Mollusks: Snails, octopi, squid, snails, clams.
- Arthropods: Insects, spiders, crustaceans such as shrimp, crabs, lobsters.

Example of Invertebrate Animals

Amoebas

Amoebas are single-celled organisms that are part of the simplest group of invertebrates. They have a cell membrane, DNA, a nucleus and organelles just like most cells in the human body. The difference is that amoebas can move using their pseudopodia, or "false feet." They also use pseudopodia to catch and eat food. Beneficial amoebas are found in soil where they help regulate bacterial populations and recycle nutrients. There are several species of amoeba that cause disease, including *Entameba hystolitica*, which causes diarrhoea.

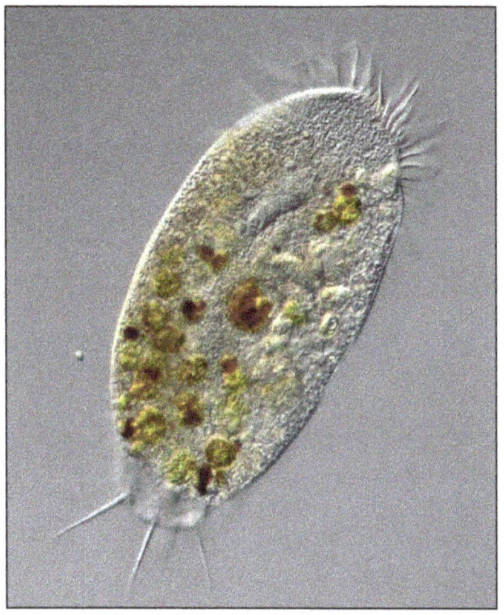

The image above shows what an amoeba looks like under a microscope.

Earthworms

Earthworms, also called angleworms, are very important in the world because their burrowing turns over and aerates the soil, provides soil drainage and mixes in organic material. They are one of more than 1,800 species of worms that live on land. Earthworms eat pieces of plants and animals in the soil. The bodies of worms are divided into segments; in certain species, some of the organs are duplicated in each segment. Earthworms can detect light and vibrations, but they are blind and cannot hear. Birds and other animals use earthworms as a food source.

Starfish

Starfish are also called sea stars because of the arms or "rays" they have sticking out of their bodies. Some starfish live deep in the oceans, while others live on the shore. There are over 1,600 species of starfish. Most have five arms, but one species is known to have 24 arms. Starfish have tube feet that work with hydraulic pressure to help them move. Snails, mussels and clams are common foods for starfish, which use their arms to guide food particles from the water into their mouths. These organisms have the unique ability to lose and regrow their arms.

The image above shows a starfish.

Squids

Squids are decapods, meaning they have 10 arms or tentacles. They are different from octopuses, which have eight arms. Each tentacle has four rows of suckers that help them catch prey (usually fish) and attach themselves to surfaces. Squids live in the ocean, both near the shore and in the ocean depths. Their length ranges from about 3/4 of an inch to more than 65 feet. Some squids are luminous, which means they have light organs in their skin that let them change color to camouflage themselves, attract prey and communicate with other squids. Humans, some fishes and sperm whales eat squid.

The image above shows a reef squid.

Spiders

Spiders make their home on every continent except Antarctica. They are not insects because they have eight legs (insects have six legs). Experts think there are about 170,000 species of spiders in the world, but only around 39,000 are described and named. Some spiders live on land and others live in the water. The leg span of species ranges from 0.02 of an inch to 10 inches. Most spiders have eight eyes, and can sense light and dark. They also are very sensitive to vibration, which comes in handy when prey gets stuck in their web. Some spiders inject poison into their prey to kill it, while others prefer to wrap up their live catch and save it for later. Spiders can only ingest liquids; they bite holes in their prey and inject enzymes that dissolve the tissues so they can be sucked out.

Animal Order

Order is more specific than class. Some of Linnaeus' orders are still used today, such as Lepidoptera (the order of butterflies and moths). There are between 19-26 orders of Mammalia, depending on how organisms are classified—sources differ. Some orders of Mammalia are Primates, Cetaceans (whales, dolphins, and porpoises), Carnivore (large carnivores/omnivores), and Chiroptera (bats).

All animals with backbones, including humans, are chordates. That is, in the language of taxonomy, they belong to the phylum Chordata. Their subphylum is Vertebrata, meaning that their backbones are segmented. Mammals, members of the class Mammalia of vertebrate animals that includes humans, are the most highly advanced organisms on Earth. They are warm-blooded, hairy, have four-chambered hearts, relatively large brains, and they suckle their young.

There are 19 orders of mammals in the world. Ten of these live in North America. Some orders include a wide range of animals; for example, shrews, lemurs, marmosets, monkeys, apes, and humans are all primates, one order of the class of mammals. Other orders are made up of only one sort of creature; Order Chiroptera, or example, consists of 18 families of bats.

The Latin names of the orders of mammals given here are followed by their common names and the families that make up each order. Examples of the various types of animals included in each family also are given.

Order Artiodactyla (Even-Toed Hoofed Animals) of Infra-Order Ungulata

Hoofed animals with an even number of toes include those that ruminate, or digest their food in four-chamber stomachs and chew cuds, and those that do not ruminate. Those that ruminate are the families Girrafidae (giraffes), Cervidae (deer, moose, reindeer, elk), Antilocapridae (pronghorn antelope) and Bovidae (cattle, bison, yaks, waterbucks, wildebeest, gazelles, springboks, sheep, musk oxen, goats). Nonruminators include the families Suidae (pigs), Tayassuidae (peccaries), Hippopotamidae (hippopotamuses), and Camelidae (camels, llamas).

Order Carnivora (Meat-Eaters)

There are two suborders of these toe-footed creatures. They include the Canidae (wolves, dogs, jackals, foxes), Ursidae (bears, giant pandas), Procyonidae (coatis, raccoons, lesser pandas), and Mustelidae (martens, weasels, skunks, otters), all part of one superfamily that is characterized by long snouts and unretractable claws; and Felidae (cats, lions, cheetahs, leopards) Hyaenidae (hyenas), and Viverridae (mongooses, civets), all of which have retractable claws. There are between 240-270 recognized species in this order, depending on the source. Some of these species have so little information on them that they are discarded by some sources. The carnivores consist of between 7-12 living families (based on source) in two superfamilies, Canoidea (or Arctoidea), the dog-like carnivores, and Feloidea (or Aeluroidea), the cat-like carnivores.

Order Cetacea (Whales and Porpoises)

Two suborders of Order Cetacea are the toothed whales, which have regular conical teeth, and the baleen, or whalebone, whales, which have irregular whalebone surfaces instead of teeth. Toothed whales include the families Physeteridae (sperm whales), Monodontidae (narwhals, belugas), Phocoenidae (porpoises), and Delphinidae (dolphins, killer whales). Baleens are in the family Eschrichtiidae (gray whales), Balaenidae (right whales), or Balaenoptridae (fin-backed whales, hump-backed whales).

Order Chiroptera (Bats)

There are two suborders of bats, the only mammals that can fly. Suborder Megachiroptera contains one family, the Pteropodidae (flying foxes, Old Worm fruit bats). Suborder Microchiroptera contains 17 families, including: Rhinopomatidae (mouse-tailed bats), Emballonuridae (sheath tailed bats), Craseonycteridae (hog-nosed or butterfly bats), Noctilionidae (bulldog or fisherman bats), Nycteridae (slit-faced bats), Megadermatidae (false vampire bats), and Rhinolophidae (horseshoe bats). 19 families, 178 genera, 926 species (the second largest mammalian order).

Order Dermoptera (Colugos or Flying Lemurs)

These gliding tree mammals from Asia do not fly and are not lemurs, but they are known as flying lemurs, or Family Cynocephalidae.

Order Edentata (Toothless Mammals)

Three families of mammals get by without teeth: Dasypodidae (armadillos), Bradypodidae (sloths), and Myrmecophagidae (hairy anteaters).

Order Hyracoidae (Hyraxes, Dassies)

Order Hyracoidae is one of three orders that has only one modern family remaining. Procavia capensis (the African rock hyrax) is one of nine living species in the Family Procaviidae.

Order Insectivora (Insect-Eaters)

The three members are the families Talpidae (moles), Soricidae (shrews), and Erinaceidae (hedgehogs).

Order Lagomorpha (Pikas, Hares, and Rabbits)

Two families make up this order: Ochotonidae (pikas) and Leporidae (hares and rabbits of all sorts).

Order Marsupialia (Pouched Animals)

Included among these are the families Caenolestidae (rat opossums), Diddeelphidae (true opossums), Dasyuridae (native cats, native mice), Notoryctidae (marsupial moles), Myrmecobiidae (numbats), Peramelidae (bandicoots), Phalangeridae (koalas), Vombatidae (wombats), and Macropodidae (kangaroos and wallabies).

Order Monotremata (Egg-Laying Mammals)

These more primitive mammals make up the families Tachyglossidae (echidnas, also called spiny anteaters) and Ornithorhynchidae (platypuses).

Order Perissodactyla (Odd-Toed Hoofed Animals) of Infra-Order Ungulata

The two suborders, Hippomorpha and Ceratomorpha, include creatures that have an odd number of toes. Families in this order are the Equidae (horses, donkeys, zebras), the Tapiridae (tapirs), and the Rhinocerotidae (rhinoceroses).

Order Pholidata

Family Manidae (pangolins) is the sole family in this order.

Order Pinnipedia (Seals and Walruses)

In the fin-footed order there are Otariidae (eared seals, sea lions), Odobenidae (walruses), and Phocidae (earless seals).

Order Primates (Primates)

The order to which people belong is divided into two suborders: The Prosimii , who have longer snouts than their relatives, and the Anthropoidae . The first group includes the families Tupalidae (tree shrew), Lemuridae (lemurs), Daubentonlidae (aye-ayes), Lorisidae (lorises, pottos), and Tarsiidae (tarsiers). The anthropoids include the families Callitrichidae (marmosets), Cebidae (New World monkeys), Cercopithecidae (baboons, Old World monkeys), Hylobatidae (gibbons), Pongidae (gorillas, chimpanzees, orangutans), and Hominidae (human beings).

Order Proboscidea (Elephants)

Large enough to have an order all to itself is Family Elephantidae.

Order Rodentia (Gnawing Mammals)

The most prolific mammals, Order Rodentia includes three suborders. It takes in the families Aplodontidae (mountain beavers), Sciuridae (chipmunks, squirrels, marmots), Cricetidae (field mice, lemmings, muskrats, hamsters, gerbils), Muridae (Old World mice, rats), Heteromyidae (New World mice), Geomyidae (gophers), and Dipodidae (jerboas), 1700 Species.

Order Sirenia (Dugongs and Manatees)

The families Trichechidae (manatees) and Dugongidae (dugongs and other sea cows) make up the Order Sirenia.

Order Tubulidentata (Aardvarks)

Another mammal in an order by itself is Family Orycteropodidae.

Animal Families

In every order, there are different animal families which all have very similar features. The Carnivora order breaks into animal families that include Felidae (Cats), Canidae (Dogs), Ursidae (Bears), and Mustelidae (Weasels).

The Dog Family (Canines)

The dog family exists almost everywhere around the world in both domestic & wild forms. This family includes wolves, foxes, and even your lovable family pet. A lot of people think canines are carnivores (eating exclusively meat), but they're actually omnivores. The omnivore diet can be easily observed by looking at the teeth, with sharp "canines" at the front, and a complete set of molars at the back (just like humans).

Dog family tracks have 4 toes, sharp claws, and their preferred style of movement is a trot. They tend to be very social, often living in large family groups that engage in cooperative hunting and parenting. Their dominant senses are primarily focused on hearing & scent, as evidenced by their large pointy ears and noses.

The Cat Family (Felines)

The cat family is another extremely common group of mammals that everyone is probably already familiar with in wild and domestic forms. In comparison to the dog family – cats are actually much more carnivorous, with their daily diet being almost 100% meat. This is why if you ever get the chance to examine cat teeth, you'll notice all their teeth are extremely sharp and pointed, even the molars.

Similar to dog tracks, cats have 4 toes, but their claws are retractable so they often do not show in the prints. Cats also tend to move much slower than the dog family. Their preferred movement style is a walk that enables them to be much stealthier. They tend to be much more solitary, with hunting & child-rearing being done solo. If you compare cat & dog faces you'll notice cat ears & noses are proportionately smaller, but their eyes are bigger. In this respect, cats are actually closer to humans and rely heavily on their sense of sight.

The Weasel Family (Mustelidae)

The weasel family contains some of the most common and abundant predators around the world, but because most of them are so small & slinky – many people have very little experience with this family. The weasel family (or mustelids) includes everything from tiny, almost mouse-sized weasels, to much larger creatures like badgers & wolverines. Weasels are incredibly vicious and efficient predators. They're capable of routinely taking down prey much larger than themselves.

They also tend to have a very high metabolism so they need to hunt and eat often. Their teeth are very sharp just like cat teeth, but perhaps their most striking physical feature is the long slinkiness of their body which enables them to enter holes and burrows in pursuit of prey. Weasel tracks have 5 toes with very sharp claws. They tend to move in a slinky loping type movement pattern, and some species like otters are excellent swimmers.

Rabbits and Hares (Leporidae)

The rabbit family is an abundant group of non-predatory animals that live in many parts of the world. Along with their cousins (the pika family), they form the Lagomorph order, which has worldwide distribution. These animals are known for having very large ears and massive hind feet they use to hop around while eating fresh vegetation.

A big part of the rabbit defence strategy from predators depends on stealth first and also their ability to run very fast when being pursued. Because rabbits are herbivores, their tooth structure is completely different from cats, dogs & weasels. They have very prominent incisors (like human front teeth) used for snipping off vegetation. A gross fact about rabbits is that they reduce the size of their digestive system by ingesting their faeces after it's already gone through once.

The Squirrel Family (Sciuridae)

Squirrels are a hugely abundant and well distributed family of rodents that can be found all over the world. Some varieties like ground squirrels live in massive colonies and burrow underground for safety, while others spend most of their time up in trees.

Rodents like squirrels are unique in the animal kingdom in that they have 4 toes on their front paws, and 5 toes on the hind. Their large front incisors can be distinguished from the teeth of other herbivores by a bright orange color. An interesting feature of the squirrel family is that many of them have an extremely well developed vocal alarm system with calls they use to warn each other of danger. Other rodent families include various families of mice, porcupines, and even larger mammals like beavers.

The Deer Family (Cervidae)

In the world of large grazing mammals, the deer family is one of the most widespread and abundant groups. This family includes a large range of sizes and habitat preferences from caribou living in open grasslands & tundra, to much smaller species living in dense jungles & forests.

Deer are ruminants, which mean they digest their fibrous diets in stages by regurgitating & chewing cud. They feed in cycles, first filling their stomach, before retreating off to safe places for rest and digestion. Like all herbivores, the deer family don't have sharp teeth for chomping meat, and instead there are just 4 large incisors at the front, and a complete set of molars at the back.

Deer tracks typically show 2 hoofed toes, except in deep substrate where their dew claws can sometimes be seen. These tracks are completely unlike any of the other mammal families. In the forest, members of the deer family can live alone or in small groups, but in more open territory it's possible to find massive herds with hundreds or even thousands of animals living together. These can be some of the largest herds in the world. The deer family tends to move in a walking pattern, but they are quite capable of jumping and quickly escaping when danger is nearby.

The Bear Family (Ursidae)

The bear family is omnivorous like canines. It includes some incredibly large animals like polar bears with adult males weighing in at 450 kg (992 lbs). While the smallest species can have adult males as small as 27 kg (60 lbs). Bears have a reputation for being dangerous because their large size & powerful jaws/claws make them very capable defenders.

They can be quite large and powerful animals, but most bears are actually much more concerned with staying away from other animals than attacking them. Most bears prefer to eat a mixed diet of fruit, berries, fish & anything else they can scavenge up in the forest, and only sometimes will actively hunt large prey.

Bear tracks have five toes with large claws used for climbing and scavenging for insects. Their hind tracks can sometimes look almost human, but the trail is unmistakably different. Bears prefer to move in a slow, ambling walk. Many species are quite skilled climbers, and they can also stand on their hind feet to get a better view. Even with their size, when provoked or scared, bears can run incredibly fast. Their dominant sense is smell as evidenced by their small eyes, small ears, and big nose.

The Great Ape Family (Hominidae)

The ape family is the most widely distributed family of mammals on the planet. Most species are found exclusively in parts of Africa & Asia, except for humans, who have managed to survive and dominate on every continent. Probably the most striking feature of apes is their incredible intelligence.

They're highly social and capable of communication skills far beyond what most other animals can demonstrate. They frequently use tools and pass on techniques for using those tools from generation to generation. Members of the ape family are technically omnivores, as evidenced by their tooth structure, which is very similar to dogs and bears. Yet many wild apes still tend to eat a mostly vegetarian diet, only occasionally dipping into meat-eating.

Animal Genus

A genus is a taxonomic category ranking used in biological classification that is below family and above species. A genus is a group of species that are closely related through common decent. A genus represents one of several hierarchical categories called taxa (singular taxongenera (plural of genus) include only a small group of species which evolved from a relatively recent common ancestor. A schematic of the overall hierarchy can be seen below for reference:

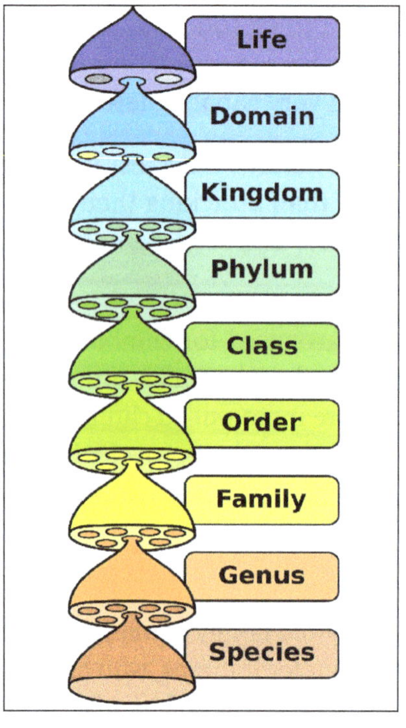

Although classification schemes vary between scientist studying different organisms, the general idea behind a genus is that it represents species that are ancestrally related, are distinct from other genera, and do not include an unnecessary amount of species. The determination of this is up to each investigator, but these general guidelines keep classifications fairly narrow when it comes to genus. The taxon above genus, Family, often includes body forms that are recognizable between groups. For instance Felidae and Canidae represent the cat-like creatures and dog-like creatures, respectively. In each family lies many genera of each type. Wolves, for instance are in the genus *Canis*, while foxes are in the genus *Vulpes*. The last step, Species, defines populations and groups that reproduce together in a continuous unit. Most of the time, different genera will not reproduce due to genetic, behavioral, or morphological differences.

Carl Linnaeus, the famous zoologist who devised a naming scheme for biological organisms (binomial nomenclature), popularized the term "genus". Although his definition of a genus varied from our modern view, the use of generic epithets in binomial nomenclature continues. The generic epithet is the first word of two in the name of an animal in binomial nomenclature. This generic epithet describes the genus that the organisms belong to. The second word or specific epithet, describes a more closely related group of organisms, or species, which the organism belongs to.

Together, these names tell you a great deal about the organisms. By knowing the genus, one also knows the family, order, and every other taxonomic classification. Because the hierarchical groups are arranged by similarities between organisms, these relationships tell us much about an individual animal. Knowing the species can inform us on how the animal is unique from others in the genus. For example, the honey badger has the scientific name: Mellivora capensis. Knowing that the honey badger is in the genus Mellivora tells us that the animal is in the family Mustelidae, meaning it is related to otters, skunks, and other small predators that are typically solitary and hunt year-round. The family Mustelidae sits inside the order Carnivora, telling us that honey badger is a carnivore, and likely hunts or scavenges for its meals. The order Carnivora sits inside the class Mammalia, telling us that the honey badger is a mammal, gives birth to live young, and feeds them with milk produced in mammary glands. Lastly, the class Mammalia sits inside the phylum Chordata and the Kingdom Animalia, telling us that the honey badger has vertebrae, bilaterally divided body plan, and is only heterotrophic.

The second name, capensis, tells us that honey badger is different from any other species in the genus Mellivora. However, the honey badger exists as the sole member of its genera, which sometimes happen. Other genera, especially of fish and invertebrates may contain hundreds of species. While it may seem that every genus should contain the same number of species, this is misleading, as the organization of species should reflect evolution. Evolutionarily, the number of organisms included in a genus should reflect the diversity of that group. If many species are being created as the genus radiates through an area, the genus will have many species. If only one species has managed to survive in a genus, the genus should contain only one species.

The organization of organisms into various taxonomic groups is a continually evolving process, and new genera are described every day. Some of these are newly found organisms, which have never had a name. Others become reclassified based on new information, such as DNA analysis. By analyzing DNA, comparing the traits of various animals, and proposing different phylogenies, scientist can hypothesize at the most likely progression of evolution. This informs the naming of various animals, and decides which groups of species can become a distinct genus.

Animal Species

A species is a group of organisms that share a genetic heritage, are able to interbreed, and to create offspring that are also fertile. Different species are separated from each other by reproductive barriers. These barriers can be geographical, such as a mountain range separating two populations, or genetic barriers that do not allow for reproduction between the two populations. Scientists have changed their definition of a species several times throughout history.

Species is one of the most specific classification that scientists use to describe animals. Scientists use a system of *binomial nomenclature* to describe animals without the confusion of common names. This system uses the *genus* as the first name, which is always capitalized, and the species name is the second name, always lower case. Thus, some animals like the Red fox, *Vulpes vulpes*, are both in the genus *Vulpes* and their species name is *vulpes*. Note the capitalization difference to distinguish between genus and species. Other foxes such as the swift fox, *Vulpes velox*, are also part of the *Vulpes*genus, but barriers exist that keep them from interbreeding with the Red foxes. In this way, they remain distinct species.

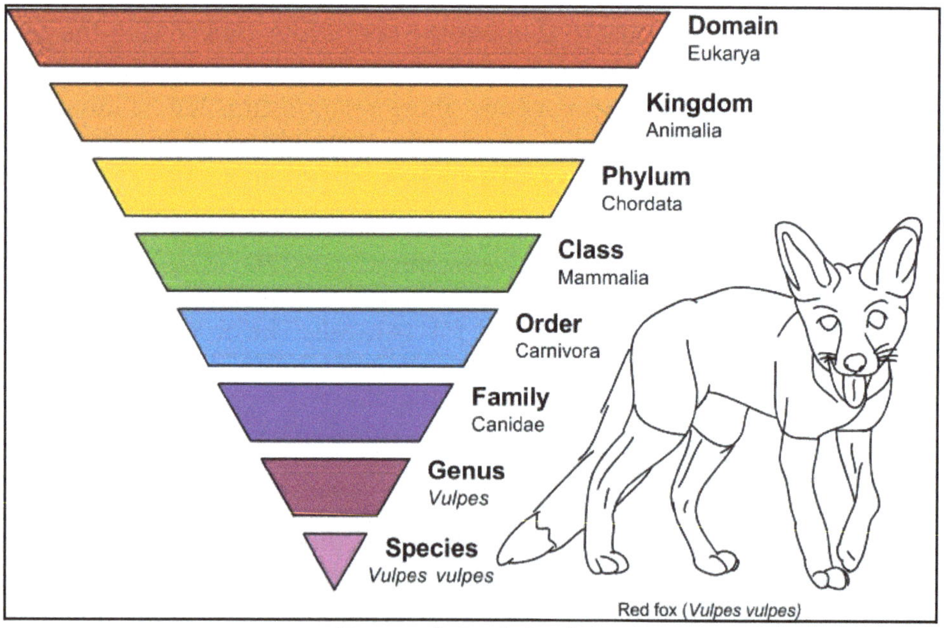

Taxonomic Rank Graph.

Since the days of Carl Linnaeus, the creator of binomial nomenclature, animals have been constantly classified and reclassified into different groups, genre, species and subspecies. Linnaeus, classifying organisms in the 1700s, restricted his classifications to the physical attributes of various organisms. Most organisms, surprisingly, he correctly identified as being related. Others, however, he couldn't have been more wrong about. Modern day genetic techniques have given us a much better window into the historical relationships between animals.

For instance, when Linnaeus first classified the elephant in the early 1700s, he only ever saw one specimen. The specimen was a fetal Asian elephant, the smallest of the known elephants today. Not knowing better, Linnaeus named the species Elephas maximus. Modern day scientists have been forced to reclassify the elephant several times. The first distinction is between the Asian and

African elephants, which are much different in size. Scientists then further had to distinguish between elephants that inhabited grasslands and those that lived in forests in Africa. Genetics show the populations do not interbreed and are separated by a reproductive barrier.

In the next century, Charles Darwin and Alfred Wallace separately conceived of the mechanism that creates multiple species from a single species. This process of natural selection applies adversity of different forms that organisms must overcome to reproduce. Organisms that are better adapted to the environment are able to reproduce more, and their offspring can also increase in number. In this way, different lines of the same species can do better or worse, depending on their genetics. Eventually, two successful lines may diverge, creating a reproductive barrier between the two populations. These populations, according to Darwin and Wallace, are now considered separate species.

Since the beginning of time, this process has been occurring and dividing organisms along different successful lineages. This theory has been confirmed by a large body of evidence. Fossil evidence provides clues that animals have been constantly changing over time, in response to a variable environment. Where Linnaeus saw animals as static, unchanging entities, it is now widely accepted that species exist on a spectrum, with some being closer related to certain species than others. Because of this, animals can often hybridize, or mate between species.

Examples of Species

Polar Bears and Grizzlies

Oftentimes the only barrier to reproduction is geographic, or based on the physical location of the animals. If this changes, the animals can interbreed, and may merge into one species. This is currently being seen in the wild in polar bears and grizzly bears. As the climate changes, polar bears are forced further south, and must start exploiting different food sources. The change in climate also allows grizzly bears to venture further north, encountering polar bears along the way. The previously separated populations now have a chance to breed, and sometimes they are successful. Hybrids have been seen in the wild, but it is not yet known if they hybrids will be successful.

There are many different situations and examples of reproductive barriers, but if the barrier can be removed, it is likely that two related species will be able to interbreed. Non-related species rarely have a possibility of breeding because they have become too different from each other. For instance, a bat and turtle have a completely different genetic makeup. The genes that control growth in the turtle would not function in a bat, and vis-versa. In fact, they don't even have the same number of chromosomes, which is a requirement for sexually reproducing organisms to be successful.

Dogs and Wolves

Still other animals, like dogs and wolves, are still technically the same species. While they have the same number of chromosomes and could technically breed, the domestic dog has come a long way from its wild counterpart. Dogs have not only evolved to be more cuddly and soft, but are attuned to human social cues. Wolves operate in a much different social structure. As such, the two are very unlikely to breed in the real world. However, because they can create fertile offspring, scientists consider them the same species.

Dogs and wolves are a good example of species radiation, or the incremental changing in a population that is widely distributed. Think of a Chihuahua. If Linnaeus had classified this animal, he certainly would not have put it in the same category as a wolf. However, a Chihuahua can breed with a slightly bigger dog, which can breed with large dog, which could easily breed with a wolf. In this way, a Chihuahua and a wolf have the same genetic basis, expressed in very different ways.

References

- Animal-kingdom, biology: byjus.com, Retrieved 13, July 2020
- Features-of-the-animal-kingdom, boundless-biology: courses.lumenlearning, Retrieved 21, January 2020
- Classification-animalia-animal-kingdom: pmfias.com, Retrieved 17, August 2020
- 5-vertebrate-groups: britannica.com, Retrieved 27, March 2020
- MammalOrders-Sheryn, WEcol-Handouts: culter.colorado.edu, Retrieved 03, June 2020
- Mammal-families: nature-mentor.com, Retrieved 19, April 2020

3

Animal Anatomy and Physiology

Animal anatomy is the study of the various forms and structures of animals. Animal physiology deals with the scientific study of the biological processes that occur for animal life to exist as well as the functioning of the animal body parts. The two subjects are closely linked and provide a better understanding of the animal. This chapter has been carefully written to provide an easy understanding of all aspects related to both animal anatomy and physiology.

Animal Anatomy

The term anatomy refers to the science that deals with the form and structure of animals. All animals are made up of cells, some of which are specialized to carry out different functions. Simple animals, such as sponges, are made up of only a few types of cell. In more complex animals, cells are organized into tissues, such as muscles and nerves that are necessary for movement. Tissues can form organs, such as the heart, which is used to pump blood around the circulatory system.

The Organs of the Body

An organ is a complex structure with a special job or a number of jobs to do. For example:

- The eye is the organ of sight.
- The kidneys are organs which get rid of water and poisonous materials from the body as urine.
- The liver has many jobs and is involved in more than one system.

Various organs are grouped together to form a body system which carries out a special job.

System of the Body	Organs in the Body	Job or Function
Musculo-skeletal	Muscle (meat), bones	Support and move the body.
Digestive	Stomach, liver, intestine, pancreas	Digest and absorb feed.
Circulatory	Heart, blood vessels	The brood carries substances around the body.
Respiratory	Muzzle, windpipe, lungs	Breathing.
Urinary	Kidneys, bladder	Get rid of poisons and waste (urine).
Nervous	Brain, nerves, spinal cord	Pass messages around the body, control the body.
Sensory	Eyes, ears, nose, skin	Sense and detect things outside the body.
Reproductive	Testes, penis ovaries, uterus, vagina, vulva, udder	To produce and feed young.
Lympho-reticular	Lymph nodes, spleen	Protect against infectious diseases, produce blood.

Animal Cells

Animal cells are typical of the eukaryotic cell, enclosed by a plasma membrane and containing a membrane-bound nucleus and organelles. Unlike the eukaryotic cells of plants and fungi, animal cells do not have a cell wall. This feature was lost in the distant past by the single-celled organisms that gave rise to the kingdom Animalia. Most cells, both animal and plant, range in size between 1 and 100 micrometers and are thus visible only with the aid of a microscope.

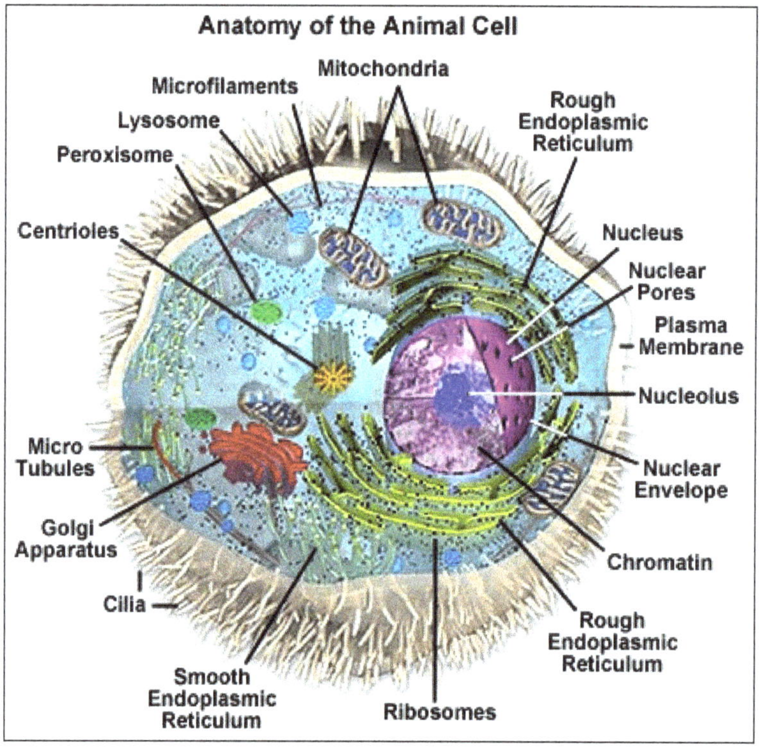

The lack of a rigid cell wall allowed animals to develop a greater diversity of cell types, tissues, and organs. Specialized cells that formed nerves and muscles—tissues impossible for plants to evolve-gave these organisms mobility. The ability to move about by the use of specialized muscle tissues is a hallmark of the animal world, though a few animals, primarily sponges, do not possess differentiated tissues. Notably, protozoans locomote, but it is only via nonmuscular means, in effect, using cilia, flagella, and pseudopodia.

The animal kingdom is unique among eukaryotic organisms because most animal tissues are bound together in an extracellular matrix by a triple helix of protein known as collagen. Plant and fungal cells are bound together in tissues or aggregations by other molecules, such as pectin. The fact that no other organisms utilize collagen in this manner is one of the indications that all animals arose from a common unicellular ancestor. Bones, shells, spicules, and other hardened structures are formed when the collagen-containing extracellular matrix between animal cells becomes calcified.

Animals are a large and incredibly diverse group of organisms. Making up about three-quarters of the species on Earth, they run the gamut from corals and jellyfish to ants, whales, elephants, and, of course, humans. Being mobile has given animals, which are capable of sensing and responding to their environment, the flexibility to adopt many different modes of feeding, defense, and reproduction. Unlike plants, however, animals are unable to manufacture their own food, and therefore, are always directly or indirectly dependent on plant life.

Most animal cells are diploid, meaning that their chromosomes exist in homologous pairs. Different chromosomal ploidies are also, however, known to occasionally occur. The proliferation of animal cells occurs in a variety of ways. In instances of sexual reproduction, the cellular process of meiosis is first necessary so that haploid daughter cells, or gametes, can be produced. Two haploid cells then fuse to form a diploid zygote, which develops into a new organism as its cells divide and multiply.

The earliest fossil evidence of animals dates from the Vendian Period (650 to 544 million years ago), with coelenterate-type creatures that left traces of their soft bodies in shallow-water sediments. The first mass extinction ended that period, but during the Cambrian Period which followed, an explosion of new forms began the evolutionary radiation that produced most of the major groups, or phyla, known today. Vertebrates (animals with backbones) are not known to have occurred until the early Ordovician Period (505 to 438 million years ago).

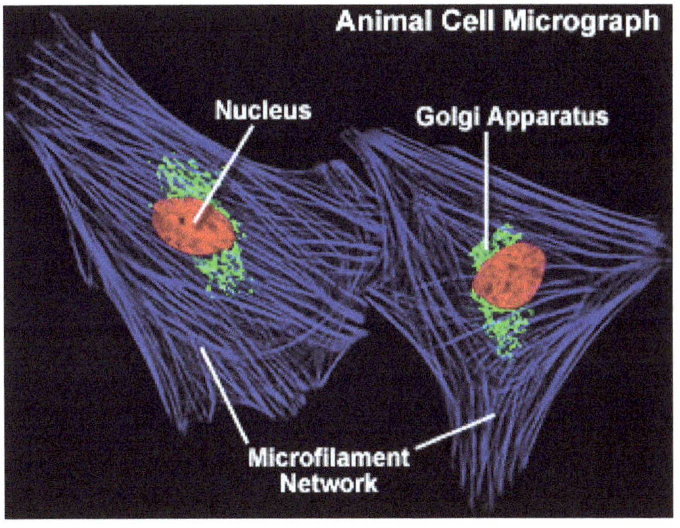

Cells were discovered in 1665 by British scientist Robert Hooke who first observed them in his crude (by today's standards) seventeenth century optical microscope. In fact, Hooke coined the term "cell", in a biological context, when he described the microscopic structure of cork like a tiny, bare room or monk's cell. Illustrated in Figure above are pair of fibroblast deer skin cells that have been labeled with fluorescent probes and photographed in the microscope to reveal their internal structure. The nuclei are stained with a red probe, while the Golgi apparatus and microfilament actin network are stained green and blue, respectively. The microscope has been a fundamental tool in the field of cell biology and is often used to observe living cells in culture. The various components that are found in animal cells:

- Centrioles: Centrioles are self-replicating organelles made up of nine bundles of microtubules and are found only in animal cells. They appear to help in organizing cell division, but aren't essential to the process.

- Cilia and Flagella: For single-celled eukaryotes, cilia and flagella are essential for the locomotion of individual organisms. In multicellular organisms, cilia function to move fluid or materials past an immobile cell as well as moving a cell or group of cells.
- Endoplasmic Reticulum: The endoplasmic reticulum is a network of sacs that manufactures, processes, and transports chemical compounds for use inside and outside of the cell. It is connected to the double-layered nuclear envelope, providing a pipeline between the nucleus and the cytoplasm.
- Endosomes and Endocytosis: Endosomes are membrane-bound vesicles, formed via a complex family of processes collectively known as endocytosis, and found in the cytoplasm of virtually every animal cell. The basic mechanism of endocytosis is the reverse of what occurs during exocytosis or cellular secretion. It involves the invagination (folding inward) of a cell's plasma membrane to surround macromolecules or other matter diffusing through the extracellular fluid.
- Golgi apparatus: The Golgi apparatus is the distribution and shipping department for the cell's chemical products. It modifies proteins and fats built in the endoplasmic reticulum and prepare them for export to the outside of the cell.
- Intermediate Filaments: Intermediate filaments are a very broad class of fibrous proteins that play an important role as both structural and functional elements of the cytoskeleton. Ranging in size from 8 to 12 nanometers, intermediate filaments function as tension-bearing elements to help maintain cell shape and rigidity.
- Lysosomes: The main function of these microbodies is digestion. Lysosomes break down cellular waste products and debris from outside the cell into simple compounds, which are transferred to the cytoplasm as new cell-building materials.
- Microfilaments: Microfilaments are solid rods made of globular proteins called actin. These filaments are primarily structural in function and are an important component of the cytoskeleton.
- Microtubules: These straight, hollow cylinders are found throughout the cytoplasm of all eukaryotic cells (prokaryotes don't have them) and carry out a variety of functions, ranging from transport to structural support.
- Mitochondria: Mitochondria are oblong shaped organelles that are found in the cytoplasm of every eukaryotic cell. In the animal cell, they are the main power generators, converting oxygen and nutrients into energy.
- Nucleus: The nucleus is a highly specialized organelle that serves as the information processing and administrative center of the cell. This organelle has two major functions: it stores the cell's hereditary material, or DNA, and it coordinates the cell's activities, which include growth, intermediary metabolism, protein synthesis, and reproduction (cell division).
- Peroxisomes: Microbodies are a diverse group of organelles that are found in the cytoplasm, roughly spherical and bound by a single membrane. There are several types of microbodies but peroxisomes are the most common.

- Plasma Membrane: All living cells have a plasma membrane that encloses their contents. In prokaryotes, the membrane is the inner layer of protection surrounded by a rigid cell wall. Eukaryotic animal cells have only the membrane to contain and protect their contents. These membranes also regulate the passage of molecules in and out of the cells.
- Ribosomes: All living cells contain ribosomes, tiny organelles composed of approximately 60 percent RNA and 40 percent protein. In eukaryotes, ribosomes are made of four strands of RNA. In prokaryotes, they consist of three strands of RNA.

Animal Tissues

The animal cells are grouped together to form animal tissues. These tissues vary in their structure, function, and origin. The animal tissues are divided into epithelial, connective, muscular and nervous tissues.

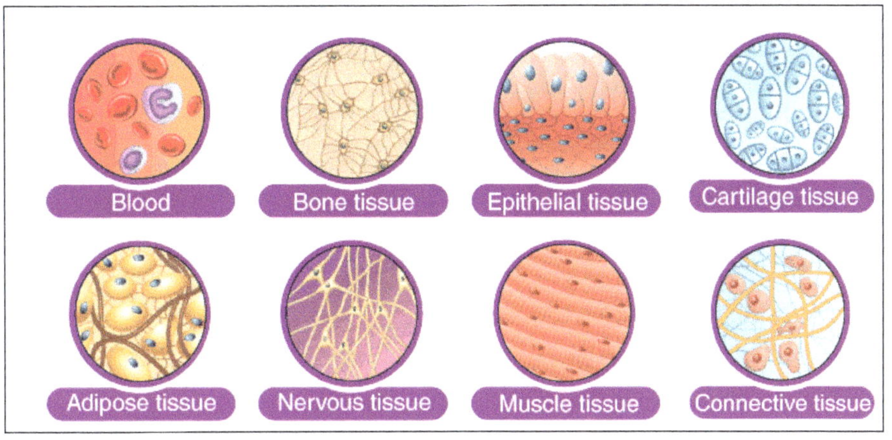

Types of Animal Tissue

The different types of animal tissues include:

Epithelial Tissue

Epithelial tissues form the protective covering and inner lining of the body and organs. These tissues were the first to evolve during evolution and were first formed during embryonic development. They develop from the ectoderm, mesoderm and endoderm of the embryo.

Characteristics of Epithelial Tissues

Following are the important characteristics of epithelial tissues:

- These can be single-layered or multi-layered.
- The tissues have the power to regenerate.
- These are held together by gap junctions, tight junctions, zonula adheren, desmosomes, or interdigitation.

- The plasma membrane of these cells is specialized into flagella, cilia, and microvilli.

Classification of Epithelial Tissues

The epithelial tissues can be classified as:

Classification	Function
Sensory epithelium	To perceive stimuli
Glandular epithelium	Secretes chemicals
Pigmented epithelium	Imparts colour in retina
Absorptive epithelium	For absorption

Connective Tissue

Connective tissues develop from the mesodermal cells of the embryo. They support and bind other tissues in the body. These are made up of three components:

- Intercellular Matrix: It is made up of mucopolysaccharide, specifically hyaluronic acid.
- Cells: The major cells include fibroblasts, adipocytes, plasma cells and mast cells.
- Fibres: Connective tissues are made up of three types of fibres, namely, collagen fibre, elastic fibre, and reticular fibre.

The connective tissues perform the following functions:

- They attach organs and tissues together.
- They store fat in the form of adipose tissues.
- They help in repairing tissues.
- They prevent the organs from mechanical shocks.
- The organs also help in defence.

Classification of Connective Tissues

The connective tissues are classified as follows:

Classification	Function
Connective Tissue Proper	Provides support and protection to the body.
Vascular Tissue	Transport of materials in the body.
Skeletal Tissue	It supports the body and gives it proper shape and form.

Muscular Tissue

The muscular tissue develops from the mesoderm of the embryo. It is classified into three types:

- Cardiac,
- Smooth,

- Skeletal.

Muscular tissue performs the following functions:

- It helps in movement and locomotion.
- It supports the bones and other structures.
- It is responsible for peristalsis and parturition.

Classification of Muscular Tissue

The muscular tissue can be classified as:

Classification	Function
Cardiac	It helps in blood circulation and keeps the heart pumping.
Smooth	These help in peristalsis and other involuntary functions of the body.
Skeletal	Provide support, help in movement and maintain homeostasis.

Nervous Tissue

Nervous tissue makes up the peripheral and the central nervous system. It develops from the ectoderm of the embryo. It possesses the ability to initiate and transmit the nerve impulse. Its main components include:

- Neurons: These are the structural and functional unit of nervous system. It comprises an axon, cell body and dendrites.
- Neuroglia: These are special cells found in the brain and spinal cord. They provide support to the neurons and fibres.
- Neuro-secretory Cells: These function as endocrine organs. They release chemical from the axons directly into blood.

Animal Physiology

Animal physiology is the scientific study of the life-supporting properties, functions and processes of animals or their parts. It focuses on how organisms, organ systems, organs, cells, and bio-molecules carry out the chemical or physical functions that exist in a living system. Therefore, the proper studying of animal physiology is crucial for understanding and evaluating underlying biological processes, behavioural states and animal response to different biological, social and environmental stimuli.

In animals, unlike humans, measurement of physiological functions is often preconditioned by a need to use invasive methods or to immobilize them by force, which induces the stress and significantly influences the outcomes of the measurement. On the other hand, anesthesia, for instance, has a profound effect on the heart or brain functions as well as precludes behavioral or mobility studies.

Circulatory System

The circulatory system is the primary method used to transport nutrients and gases through the body. Simple diffusion allows some water, nutrient, waste, and gas exchange in animals that are only a few cell layers thick; however, bulk flow is the only method by which the entire body of larger, more complex organisms is accessed.

Circulatory System Architecture

The circulatory system is effectively a network of cylindrical vessels: the arteries, veins, and capillaries that emanate from a pump, the heart. In all vertebrate organisms, as well as some invertebrates, this is a closed-loop system, in which the blood is not free in a cavity. In a closed circulatory system, blood is contained inside blood vessels and circulates uni-directionally from the heart around the systemic circulatory route, then returns to the heart again.

As opposed to a closed system, arthropods– including insects, crustaceans, and most mollusks– have an 'open' circulatory system. In an open circulatory system, the blood is not enclosed in blood vessels but is pumped into an open cavity called a hemocoel and is called hemolymph because the blood mixes with the interstitial fluid. As the heart beats and the animal moves, the hemolymph circulates around the organs within the body cavity and then re-enters the hearts through openings called ostia. This movement allows for nutrient exchange, and in some organisms lacking direct gas exchange sites, a basic mechanism to transport gasses beyond the exchange site. Because the gas exchange in many open-circulatory systems tends to be relatively low for metabolically-active organs and tissues, a trade-off exists between this system and the much more energy-consuming, harder-to-maintain closed system.

Digestive System

Animals, for the most part, ingest their food as large, complex molecules that must be broken down into smaller molecules (monomers) that can then be distributed throughout the body of every cell. This vital function is accomplished by a series of specialized organs that comprise the digestive system. Animals use the organs of their digestive systems to extract important nutrients from food they consume, which can later be absorbed.

Digestion is the process of breaking down feed into simple substances that can be absorbed by the body. Absorption is the taking of the digested parts of the feed into the bloodstream. The digestive system consists of the parts of the body involved in chewing and digesting feed. This system also moves the digested feed through the animal's body and absorbs the products of digestion. Different species of animals are better able to digest certain types of feeds than others. This difference occurs because of the various types of digestive systems found in animals. There are four basic types of digestive systems: monogastric, avian, ruminant, and pseudo-ruminant.

Monogastric Digestive System

A monogastric digestive system has one simple stomach. The stomach secretes acid, resulting in a low pH of 1.5 to 2.5. The low pH destroys most bacteria and begins to break down the feed materials. Animals with this type of digestive system are better adapted to eat rations high in concentrates.

Concentrates are highly digestible feedstuffs that are high in energy and low in fiber. Concentrates are typically 80 to 90 percent digestible. Common concentrates are cereal grains and oil meals. Cereal grains include corn, wheat, barley, and oats. Oil meals include soybean meal, linseed meal, and cottonseed meal. Examples of monogastric animals are hogs, cats, dogs, and humans.

Excretory System

Microorganisms and invertebrate animals use more primitive and simple mechanisms to get rid of their metabolic wastes than the mammalian system of kidney and urinary function. Three excretory systems evolved in organisms before complex kidneys: vacuoles, flame cells, and Malpighian tubules.

Contractile Vacuoles in Microorganisms

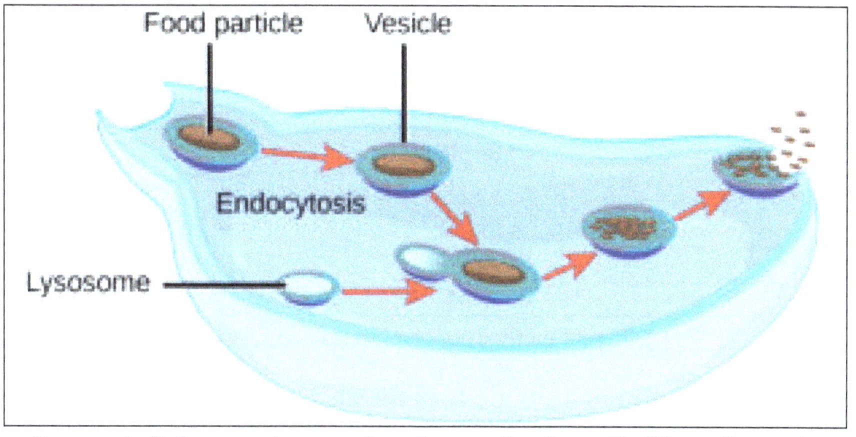

Some unicellular organisms, such as the amoeba, ingest food by endocytosis. The food vesicle fuses with a lysosome, which digests the food. Waste is excreted by exocytosis.

The most fundamental feature of life is the presence of a cell. In other words, a cell is the simplest functional unit of a life. Bacteria are unicellular, prokaryotic organisms that have some of the least complex life processes in place; however, prokaryotes such as bacteria do not contain membrane-bound vacuoles. The cells of microorganisms like bacteria, protozoa, and fungi are bound by cell membranes and use them to interact with the environment. Some cells, including some leucocytes in humans, are able to engulf food by endocytosis—the formation of vesicles by involution of the cell membrane within the cells. The same vesicles are able to interact and exchange metabolites with the intracellular environment. In some unicellular eukaryotic organisms such as the amoeba, shown in Figure, cellular wastes and excess water are excreted by exocytosis, when the contractile vacuoles merge with the cell membrane and expel wastes into the environment. Contractile vacuoles (CV) should not be confused with vacuoles, which store food or water.

Respiratory System

Respiratory system the system in living organisms that takes up oxygen and discharges carbon dioxide in order to satisfy energy requirements. In the living organism, energy is liberated, along with carbon dioxide, through the oxidation of molecules containing carbon. The term respiration denotes the exchange of the respiratory gases (oxygen and carbon dioxide) between the organism and the medium in which it lives and between the cells of the body and the tissue fluid that bathes them.

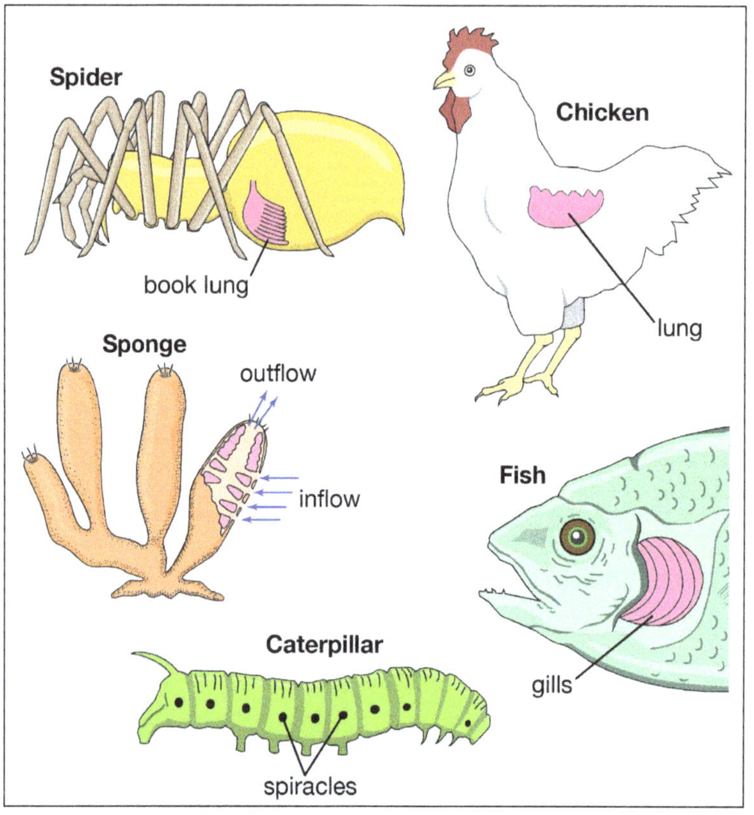

Different methods of respiration in animals.

With the exception of energy used by animal life in the deep ocean, all energy used by animals is ultimately derived from the energy of sunlight. The carbon dioxide in the atmosphere in conjunction with the energy of sunlight is used by plants to synthesize sugars and other components. Animals consume plants or other organic material to obtain chemical compounds, which are then oxidized to sustain vital processes.

Although the acquisition of oxygen and the elimination of carbon dioxide are essential requirements for all animals, the rate and amount of gaseous exchange vary according to the kind of animal and its state of activity. The oxygen consumption of various animals is expressed in terms of millilitres of oxygen per kilogram of body weight per hour, reflecting the gas demands of different species at rest and in motion. A change in the chemical composition of the body fluids elicits a response from the central nervous system, which then excites or depresses the machinery of external respiration.

Basic Types of Respiratory Structures

Respiratory structures are tailored to the need for oxygen. Minute life-forms, such as protozoans, exchange oxygen and carbon dioxide across their entire surfaces. Multicellular organisms, in which diffusion distances are longer, generally resort to other strategies. Aquatic worms, for example, lengthen and flatten their bodies to refresh the external medium at their surfaces. Sessile sponges rely on the ebb and flow of ambient water. By contrast, the jellyfish, which can be quite large, has a low oxygen need because its content of organic matter is less than 1 percent and its metabolizing cells are located just beneath the surface, so that diffusing distances are small.

Organisms too large to satisfy their oxygen needs from the environment by diffusion are equipped with special respiratory structures in the form of gills, lungs, specialized areas of the intestine or pharynx (in certain fishes), or tracheae (air tubes penetrating the body wall, as in insects).

Respiratory structures typically have an attenuated shape and a semipermeable surface that is large in relation to the volume of the structure. Within them there is usually a circulation of body fluids (blood through the lungs, for example). Two sorts of pumping mechanisms are frequently encountered: one to renew the external oxygen-containing medium, the other to ensure circulation of the body fluids through the respiratory structure. In air-breathing vertebrates, alternately contracting sets of muscles create the pressure differences needed to expand or deflate the lungs, while the heart pumps blood through the respiratory surfaces within the lungs. Oxygenated blood returning to the heart is then pumped through the vascular system to the various tissues where the oxygen is consumed.

Nervous System

The nervous system is the part of an animal's body that coordinates its behavior and transmits signals between different body areas. In vertebrates it consists of two main parts, called the central nervous system (CNS) and the peripheral nervous system (PNS). The CNS contains the brain and spinal cord. The PNS consists mainly of nerves, which are long fibers that connect the CNS to every other part of the body, but also includes other components such as peripheral ganglia, sympathetic and parasympathetic ganglia, and the enteric nervous system, a semi-independent part of the nervous system whose function is to control the gastrointestinal system.

Neuron Structure

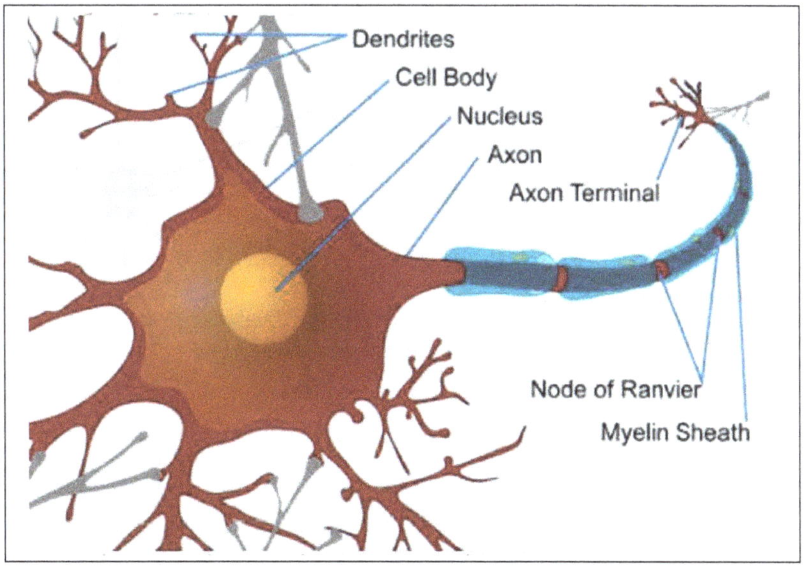

The structure of a neuron allows it to rapidly transmit nerve impulses to other cells.

A neuron consists of three basic parts: the cell body, dendrites, and axon:

- The cell body contains the nucleus and other cell organelles.
- Dendrites extend from the cell body and receive nerve impulses from other neurons.

- The axon is a long extension of the cell body that transmits nerve impulses to other cells. The axon branches at the end, forming axon terminals. These are the points where the neuron communicates with other cells.

Nervous System Signal move from One Cell to the Next

It literally jumps by way of a chemical transmitter. Notice the two cells are not connected, but separated by a small gap known as synapse. It is the space between a neuron and the next cell.

Nerve Impulses

Nerve impulses are electrical in nature. They result from a difference in electrical charge across the plasma membrane of a neuron. How does this difference in electrical charge come about? The answer involves ions, which are electrically charged atoms or molecules.

The Synapse

The place where an axon terminal meets another cell is called a synapse. The axon terminal and other cell are separated by a narrow space known as a synaptic cleft. When an action potential reaches the axon terminal, the axon terminal releases molecules of a chemical called a neurotransmitter. The neurotransmitter molecules travel across the synaptic cleft and bind to receptors on the membrane of the other cell. If the other cell is a neuron, this starts an action potential in the other cell.

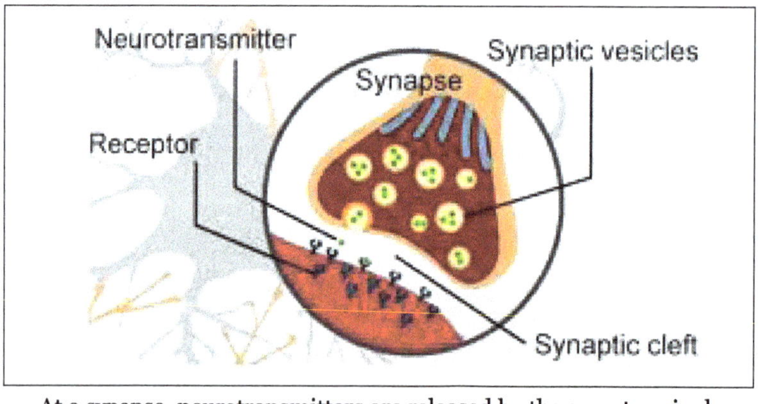

At a synapse, neurotransmitters are released by the axon terminal. They bind with receptors on the other cell.

Endocrine System

The endocrine system of animals encompasses a group of tissues that release hormones into circulation for travel to and action on distant targets. An endocrine tissue is typically a ductless gland (e.g, pituitary, and thyroid) that releases its hormones into capillaries that permeate the tissue. These glands are richly supplied with blood. However, non-typical endocrine tissues also contribute important hormones to circulation, e.g, and secretion of atrial natriuretic peptide from the heart, erythropoietin from the kidney, insulin-like growth factor from the liver, and leptin from fat. New hormones continue to be discovered. Some act only on a single tissue, whereas others have effects on virtually all cells of the body. The effects of hormones on their targets are varied—from enhancement of nutrient uptake to altering cell division and differentiation, among many others. Considered to be part of animal's communication system is Nervous system uses physical structures for communication and Endocrine system uses body fluids to transport messages (hormones) which referred to as humeral versus neural control.

Hormones

Classically, hormones are defined as chemical substances produced by ductless glands and secreted into the blood supply to affect a tissue distant from the gland, but now it is understood that hormones can be produced by single cells as well:

- Epicrine are hormones pass through gap junctions of adjacent cells without entering extracellular fluid.
- Neurocrine are hormones pass through synaptic clefts between neurons as do neurotransmitters; hormones can also be synthesized in a neuron (e.g. oxytocin) and secreted directly into the blood from the axon.
- Paracrine are hormones diffuse through interstitial fluid (e.g. prostaglandins).
- Endocrine are hormones are delivered via the bloodstream (e.g. growth hormone).
- Exocrine are chemical substance is secreted to exterior of the body (e.g. phermones) or into the digestive tract (e.g. somatostatin).

In Biochemistry, Hormones are biochemically categorized as amines, peptides, steroids, or prostaglandins:

- Amines are thyroid hormones, adrenal catecholamines.
- Peptides are growth hormone, prolactin, luteinizing hormone.
- Steroids are estradiol, testosterone, cortisol, progesterone.
- Prostaglandins are PGF2.

Pituitary gland (hypophysis cerebri) has two distinct parts that are anterior lobe (adenohypophysis) and posterior lobe (neurohypophysis) which is located in a bony recess (sella turcica) at the base of the brain that are connected to the brain by the hypothalamus and a portal blood supply, vein draining the hypothalamus breaks up into a capillary bed within the anterior pituitary and route by which releasing factors from the hypothalamus travel to cause release of hormones from the anterior pituitary.

Figure shows schematic representation of the pituitary. 1. Hypothalamic-hypophyseal portal circulation and the pathway followed by the hypothalamic-hypophysiotropic hormones (releasing and inhibiting) in reaching the anterior pituitary (left). 2. Pathway by which the posterior pituitary hormone oxytocin and ADH (antidiuretic hormone) reach the capillaries of the posterior lobe (right). Hormone-containing blood leaves each lobe of the pituitary through a number of hypophyseal veins. The orange arrows indicate the direction of blood flow; the black arrowheads indicate the path taken by the hormones involved.

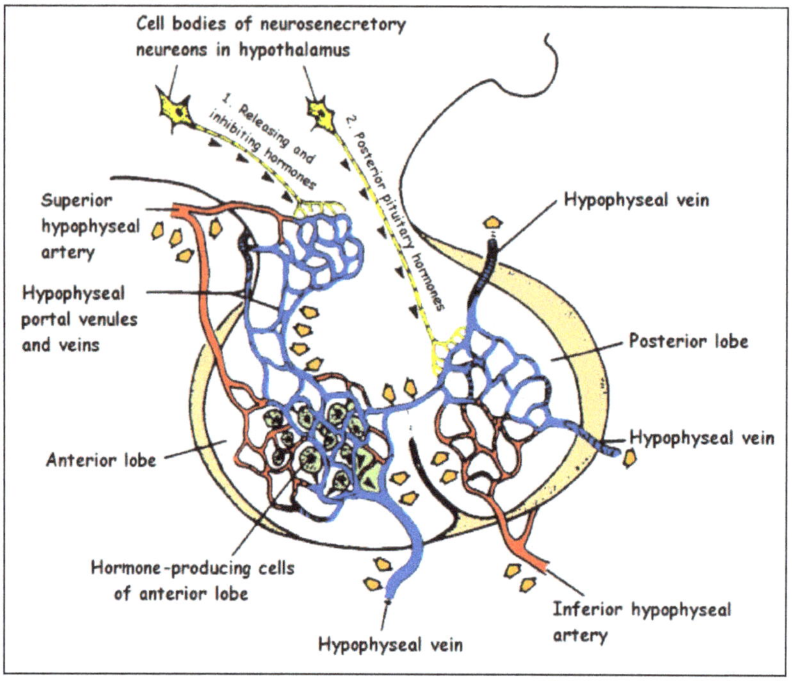

Posterior pituitary is an outgrowth of the hypothalamus and contains the terminal axons from two pairs of nuclei that is supraoptic that secrets antidiuretic hormone and paraventricular which secretes oxytocin. The action potentials generated by the need for each of the stored hormones causes release of the substance into the blood supply.

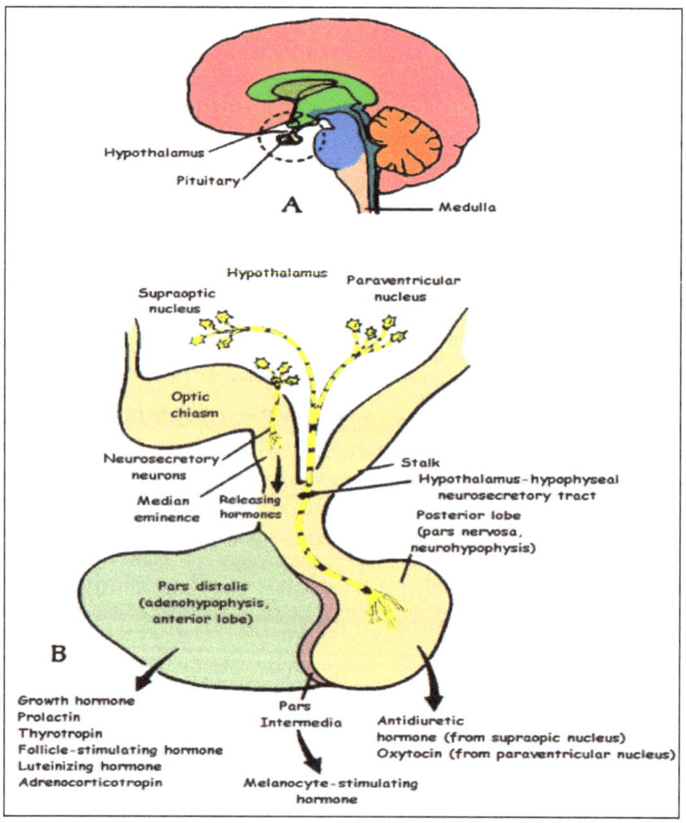

Location of the pituitary in the primate brain. A. The pituitary is situated at the base of the hypothalamus. B. Hypothalamic neurons from the supraoptic and paraventricular nuclea transmit their neurosecretions of antidiuretic hormone and oxytocin, respectively, to the posterior lobe.

Anterior Pituitary

Anterior pituitary lies forward from the posterior pituitary and is composed of three parts:

- Pars distalis: It has principal part – 5 cell types that are:
  - Somatotropes: Growth hormone.
  - Corticotropes: Corticotropin releasing hormone.
  - Mammotropes: Prolactin.
- Pars intermedia:
  - Melanocyte stimulating hormone.
- Pars tuberalis is layer of cells surrounding the neural stalk with no know function.

Functions of Anterior Pituitary Hormones

The main functions of anterior pituitary hormones are growth hormone that regulates growth, metabolism and acts on liver to cause production of somatomedins and somatomedin binding proteins.

Adrenocorticotrophic hormone stimulates adrenal cortex to release cortisol and aldosterone, involved in metabolic regulation. Gonadotrophic hormones: follicle stimulating hormone which stimulates follicular growth in the ovary and luteinizing hormone which stimulates ovulation and growth of corpora lutea.

Thyroid stimulating hormone which stimulates release of thyroxine from thyroid gland, prolactin which stimulates mammary development and lactation and beta-lipotrophin, secreted by corticotropes and involved in pain relief. Functions of posterior pituitary hormones are ADH – regulates reabsorption of water in the kidney and Oxytocin – milk ejection reflex.

Thyroid Gland

In most mammals the thyroid gland is located on the trachea, just caudal to the larynx.

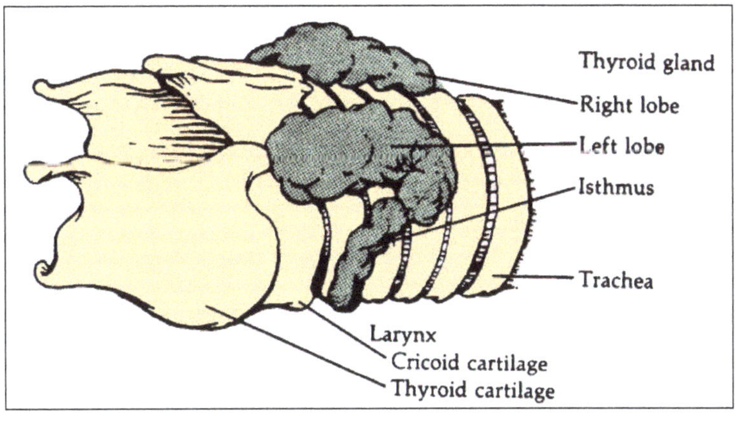

Thyroid Hormones

Thyroid hormones belong to the amine classification of hormones. They are derived from the amino acid tyrosine. Thyroid hormones also contain iodine.

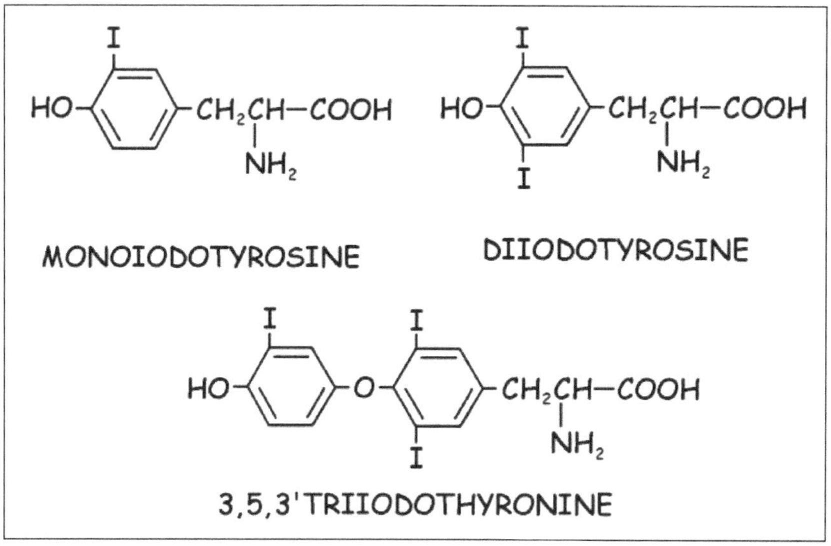

Structural formulas of the iodinated amino acids monoiodotyrosine and diiodotyrosine and the compounds formed by their coupling. The coupling of monoiodotyrosine and diiodotyrosine with loss of H_2O and glycine produces 2,5,3'-triiodotyrosine, known as T_3. The coupling of two molecules of diiodotyrosine with loss of H_2O and glycine produces thyroxine, known as T_4.

Thyroid hormones increase basal metabolism and heat production. Secretion of thyroid hormones (thyroxine and triiodothyronine) is regulated by thyroid stimulating hormone (TSH) from the anterior pituitary. Thyroglobulin is a large glyporprotein molecule secreted into the thyroid follicles by the cells lining the follicles. Thyroid deficiency typically results from iodine deficiency. The thyroid gland enlarges because of accumulation of thyroglobulin. Hypo and hyper thyroidism are rare in domestic animals, bud it does occur in dogs and cats. Calcitonin is the hormone of the thyroid gland secreted by the parafollicular or C cells and inhibits calcium reabsorption at the kidney and stimulates calcium deposition in bone – lowers blood calcium.

Parathyroid Gland

One or two glands which are located near or embedded in the thyroid glands. Low calcium levels in plasma stimulate parathyroid hormone (PTH) release. PTH stimulates the release of calcium and phosphorus from the bones. PTH also stimulates osteoclasts to dissolve bone. PTH enhances both calcium and phosphate absorption from the intestine by increasing the rate of formation of 1,25-dihydroxycholecalciferol, the active form of vitamin D.

Adrenal Glands

Hormones from the adrenal gland are steroids formed mostly from cholesterol. Adrenal glands are small, paired structures that lie immediately cranial to the kidneys and are close to the junction of the renal vein with the posterior vena cava.

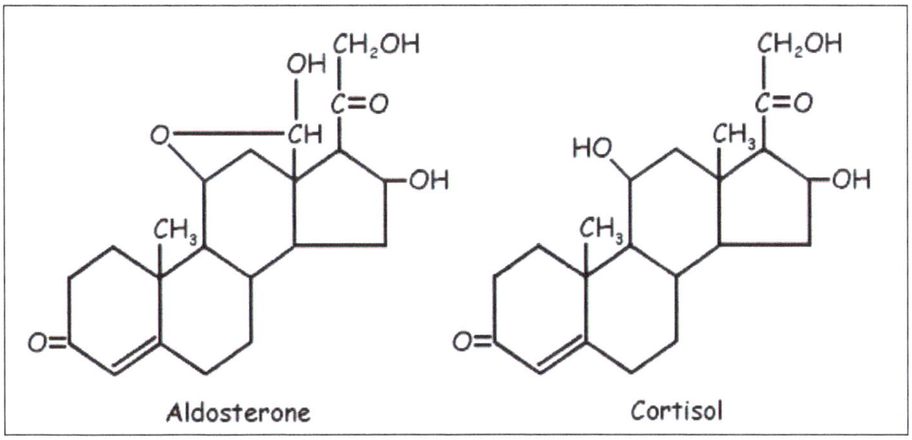

Structural formulas of the principal adrenocortical hormones.

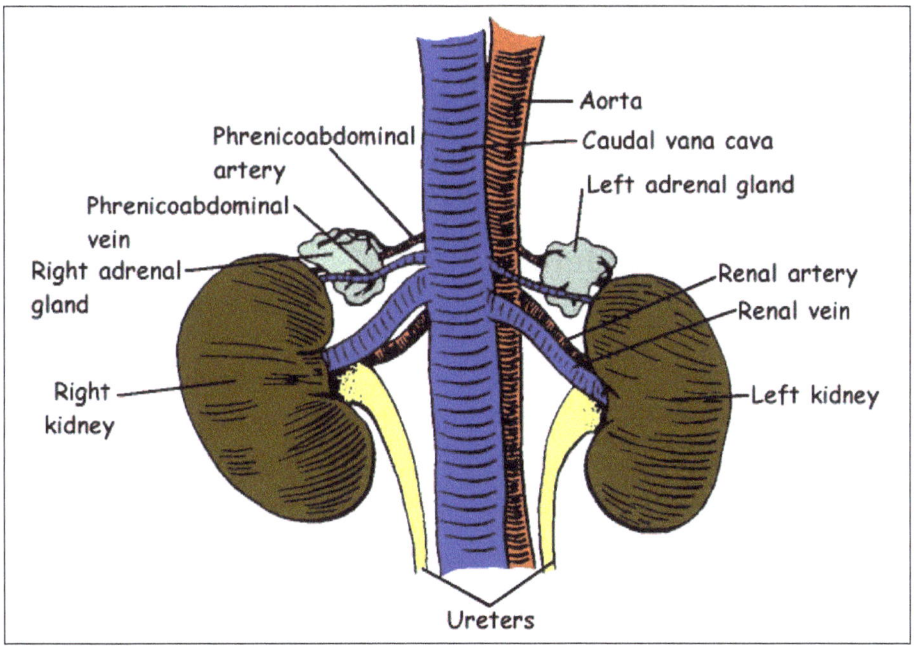

The canine adrenal glands (ventral view). Their blood supply and venous drainage is by way of branches from the phreniocabdominal arteries and veins.

Seven adrenocortical hormones are recognized as secretions of the adrenal cortex. Corticosterone, cortisone, 11-dehydroxycorticosterone, are called glucocorticoids. It plays major role in carbohydrate metabolism. It enhances gluconeogenesis, production of glucose from non- carbohydrate precursors, enhance water and sodium loss, also are potent anti-inflammatory compounds. Glucocorticoids secreted by the zona fasiculata of the adrenal cortex and secretion of glucocorticoids regulated by ACTH from pituitary gland.

11-deoxycorticosterone, 17-hydroxy-11deoxycorticosterone, and aldosterone are all called mineral corticoids. It enhances sodium retention and potassium excretion, mineral corticoids are secreted from zona glomerulosa of adrenal cortex. The regulation of aldosterone secretion is under the rennin-angiotensin system. Rennin is secreted by juxtaglomerular cells in kidney and acts on angiotensinogen in blood to form angiotensin I. Angiotensin I is converted in the lungs to angiotensin II which causes aldosterone release.

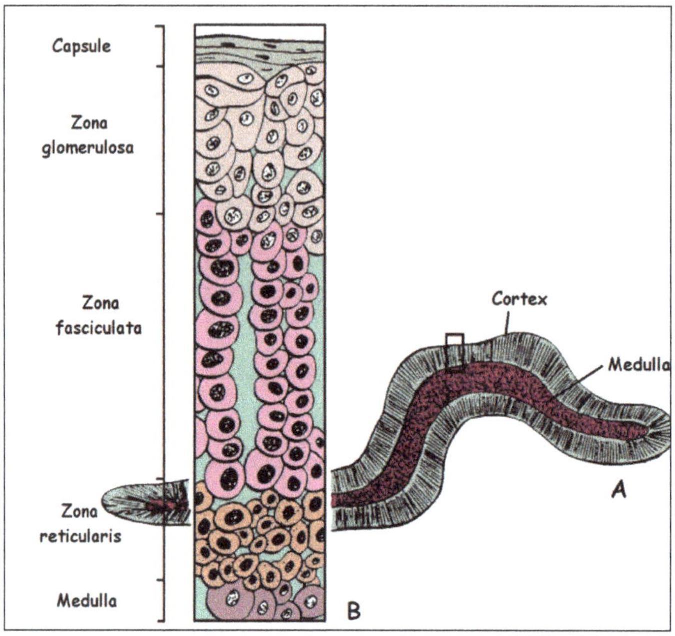

Contrasting appearance of the cortex and medulla. B. Magnification of boxed-in area in A that shows the different cell types associated with the three zones of the cortex.

Hormones of the Adrenal Medulla

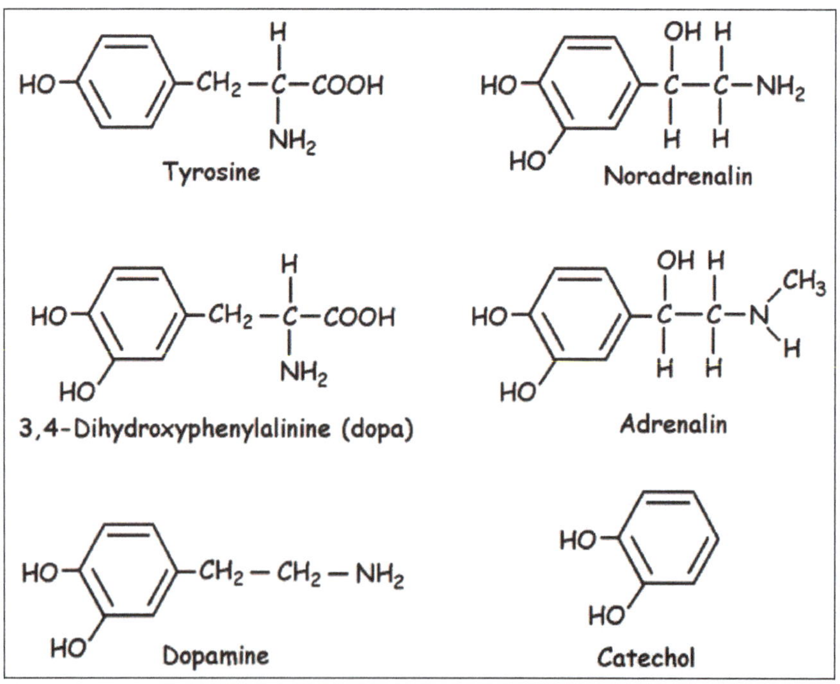

Tyrosine and are derivatives of catechol. The abbreviation "dop" is derived from the German name of this compound, dioxyphenylalamine.

Hormones of the adrenal medulla belong to the amine class and are known as epinephrine and nor-epinephrine. These hormones are referred to as catecholamines and are derived from the amino acid tyrosine. Epinephrine is secreted only by the adrenal medulla, but norepinephrine is also secreted by the post-ganglionic sympathetic neurons. Catecholamines have pronounced metabolic effects. Catecholamines are also associated with the fight or flight response. Two types of receptors are associated with binding catecholamines. These are alpha receptors and beta receptors. Alpha receptor is stimulatory, except in intestinal smooth muscle where they are inhibitory. Beta receptor is inhibitory, except in cardiac muscle where they are stimulatory.

Pancreas

Hormones of the pancreas are insulin, glucagon, somatostatin, and pancreatic polypeptide. Each hormone is secreted by a specific cell type in the pancreatic islets:

- Alpha cells: Glucagon which stimulates production of glucose from liver.
- Beta cells: Insulin which stimulate uptake of glucose.
- Delta cells: Somatostatin which inhibits secretion of insulin and glucagon inhibits secretion of gastrin, secretin, and gastric acid.
- F cells: pancreatic polypeptide. The secretion is stimulated by protein intake, exercise, and fast-ing – function unknown.

Prostaglandins and their Functions

First isolated from prostate gland and hence named prostaglandins which derived from arachidonic acid. Prostaglandins promote inflammation and are involved in blood clotting. Prostaglandins are also involved in the regulation of ovarian function.

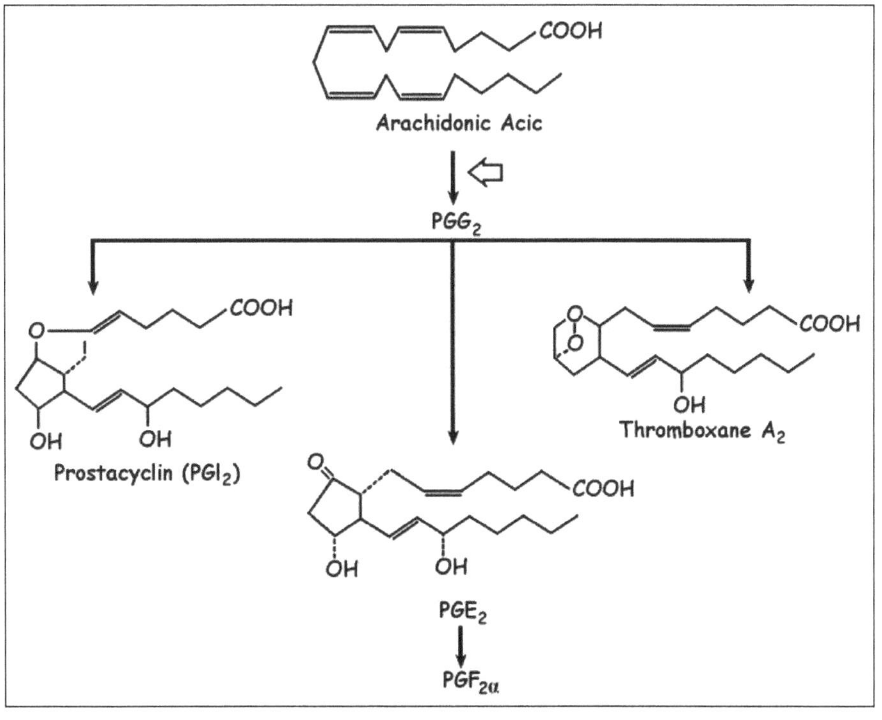

Thromboxane A2 is biochemically related to the prostaglandins and is formed from them as shown. Thromboxane A2 promotes the platelet release reaction associated with blood coagulation. Therefore, aspirin retards blood coagulation.

Animal Locomotion

Locomotion is an important process for animals. Animals have to move from one place to another for many reasons. They can't stay in one place in order to support their living. Hence, certain body movements are observed in every organism but the means vary according to their body structure. The gait of a few animals is as follows:

Earthworm

Locomotion of earthworm is the outcome of muscular actions in them. Continuous expansion and contraction of muscles help them to move. Hence, earthworms make small moves using muscles alone.

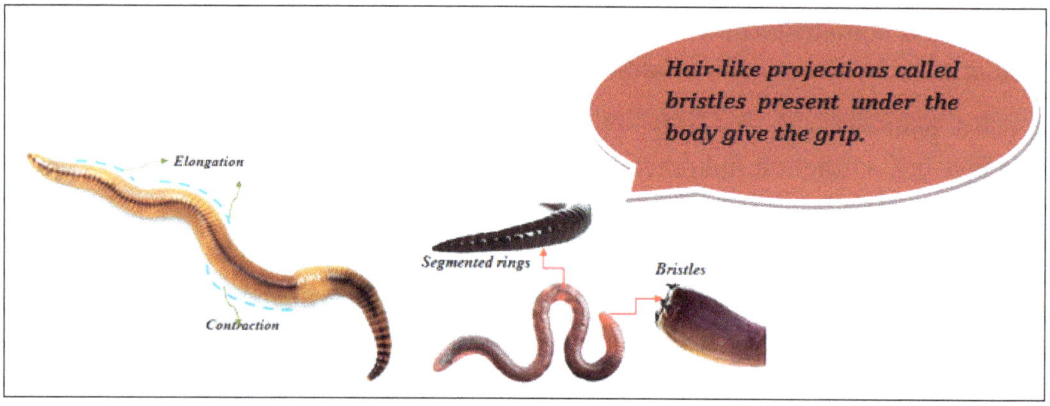

Locomotion in Earthworm.

Snail

Snails have a characteristic shell on their back. The hard external shell is a non-bony structure called exoskeleton. It does not help in movements but is a shield from predators and damages.

Cockroach

Cockroaches have a body structure that facilitates walking, climbing and flying. Their body is covered by a hard outer skeleton.

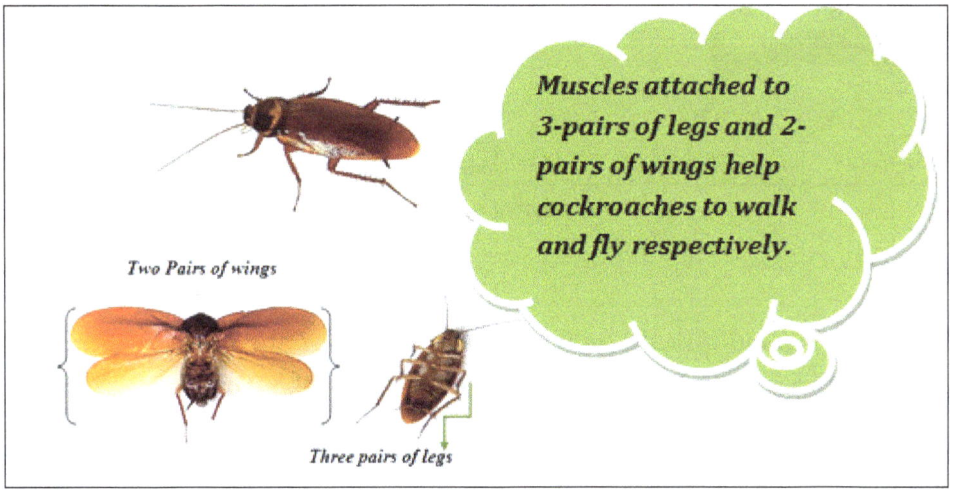

Birds

Birds have light bones attached to strong muscles which help them to fly. Whereas hind limb bones are modified to walk and perch.

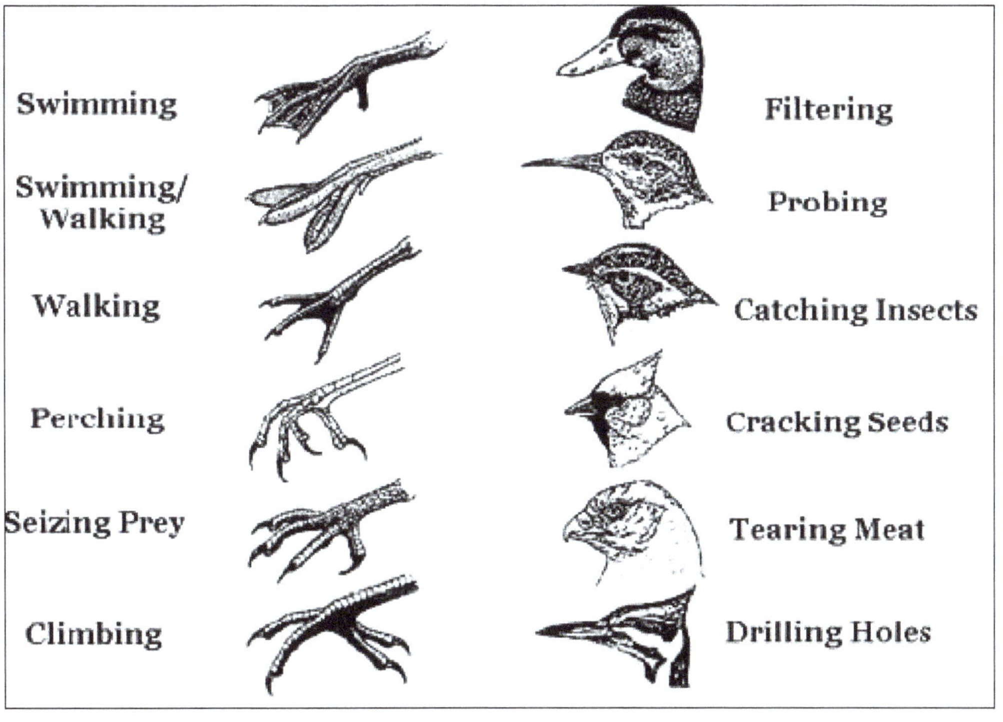

Fish

Fish is an aquatic animal adapted to live in water.

Locomotion in Fish.

Snakes

Snakes slither on the ground with the help of its long backbone and muscles attached to it.

Locomotion in Snakes.

Slither, glide, hop, wiggle, run, jump – the list of ways animals move is countless! Animals of all size and shape move in search of shelter, food, mate and to escape from predators using their different body parts like fins, legs, wings and so on.

Thermoregulation

Animals occupy diverse types of biomes and aquatic environments. A key factor in their success is the ability for many species to adapt to external environmental fluctuations and maintain a consistent internal environment. Cannon described the capacity for an animal to regulate its internal environment as the product of a suite of physiological processes, called homeostasis. Homeostatic mechanisms are dynamic and regulate many different parameters in the animal body (e.g., pH, dissolved oxygen, glucose concentration). Internal temperature changes may adversely affect many aspects of animal physiology, including enzyme function, muscle activity, and energy metabolism.

Types of Thermoregulation

There are two primary responses to fluctuating ambient temperatures (T_A) exhibited by animals: poikilothermy and homeothermy. Because poikilotherms lack the physiological means to generate heat, the body temperature of these animals tends to conform to that of the outside environment in the absence of any behavioral intervention. Examples of poikilotherms include the "cold-blooded" animals, such as most fish, amphibians, and reptiles. On the other hand, homeotherms have specific physiological adaptations for regulating their body temperatures; body temperatures of homeotherms do not fluctuate as much as those of poikilotherms. Indeed, all homeotherms maintain high body temperatures in the range of 36 to 42 °C and include the "warm-blooded" animals, such as birds and mammals.

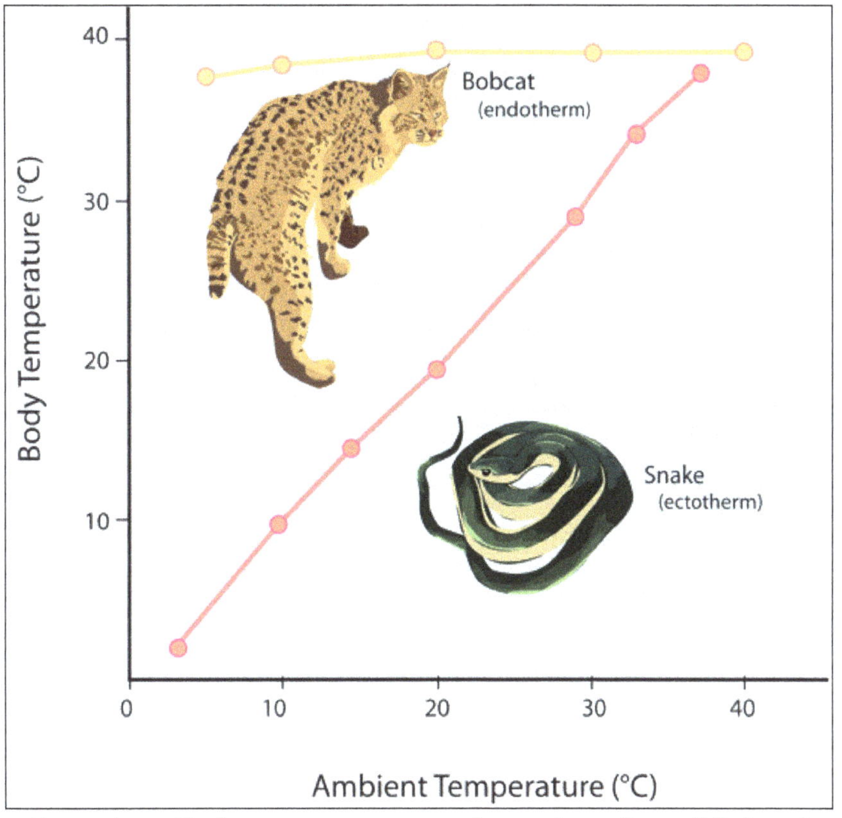

Comparison of body temperature response by ectotherm (i.e., poikilotherm) and endotherm (i.e., homeotherm) to changing ambient temperatures.

Poikilotherms are also known as ectotherms because their body heat is derived exclusively from their external environments. This external thermal dependence enables them to employ behavioral thermoregulation by 1) shuttling between areas with lower and higher temperatures and 2) changing body positions to adjust heat exchange via conduction and radiation. For example, wood turtles (*Glyptemys insculpta*) move daily into forest clearings to bask and elevate their body temperatures but return to streams at night because the water temperature does not drop as much as does the air temperature. Behavioral thermoregulation in poikilotherms can result in sustained average body temperatures that are optimal for crucial metabolic activities, such as feeding and locomotion. Reiserer et al describe groups of neonate sidewinder rattlesnakes (*Crotalus cerastes*) that aggregate as "ball" groups during the time period leading up to the first shedding. The ball core temperature remained stable (mean of 31.94 ± 2 °C) over a 5-hour period during the hottest part of the day, while temperatures measured outside the balls exceeded 42 °C, the lethal critical maximum for these snakes.

Melanistic polymorphism among related species of lizards may be related to thermoregulation. In cordylid (*Cordylus*) lizards, melanistic species warm up more quickly than do lighter-colored species with higher thermal reflectance. This heat retention helps these lizards remain active under cold conditions. Other kinds of lizards, such as Amphibolurus decresii and Anolis carolinensis, exhibit labile body coloration but the color choice is not related to thermal condition but, instead, functions in social or camouflage contexts. For example, the agamid lizard Pseudotrapelus sinaitus uses body color changes to signal to conspecifics and not to thermoregulate.

Homeotherms also use behavioral thermoregulation (i.e., habitat choice) to adjust their body temperatures but, unlike poikilotherms, they do not depend solely on the outside environment as a source of body heat. Instead, homeotherms use physiological mechanisms to regulate their body temperatures independently from ambient temperatures. When low ambient temperatures threaten to overcool the regulated body temperature, homeotherms have several strategies to supplement and conserve body heat. Isometric contraction of skeletal muscles, called shivering, transfers mechanical heat to the body core while vasoconstriction of peripheral vessels reduces heat loss from the integument. Heinrich described how shivering by bumblebees in cold ambient temperatures increased thoracic body temperatures to the level required for flight muscle activity (i.e., 30–40 °C). In addition, some homeotherms are adapted for non-shivering thermogenesis, a metabolic process in which brown adipose tissue is catabolized not for ATP synthesis, but for heat production. As ambient temperatures increase, homeotherms use evaporative cooling through sweating and/or panting to regulate body temperatures, and also vasodilate surface blood vessels to promote heat loss.

Thermoneutral Zone

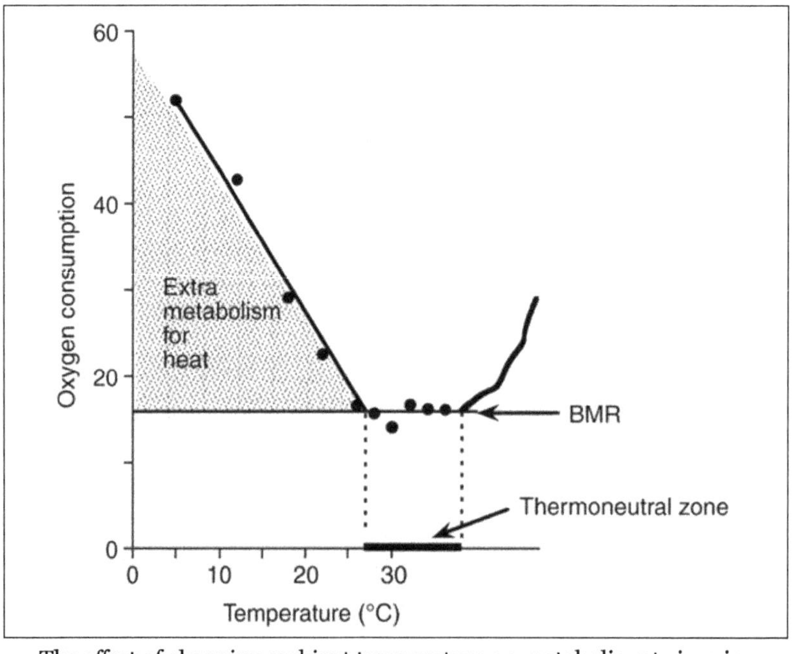

The effect of changing ambient temperature on metabolic rate in mice above and below the thermoneutral zone. BMR is the basal metabolic rate.

A homeotherm's basic needs are met through basal metabolism, a form of metabolism that does not involve physiological thermoregulation because external temperatures do not exceed comfortable limits. The range of temperatures associated with basal metabolism comprises the thermoneutral zone. When ambient temperatures either exceed or fall below the thermoneutral zone, physiological strategies like those are deployed to prevent the body temperature from changing. However, whenever an animal is thermoregulating to prevent overheating or overcooling, the non-basal metabolic rate increases. This increase in metabolic rate constrains the limits of what temperatures can be tolerated beyond the thermoneutral zone, especially the upper temperature limit. In general, homeotherms utilize behavioral means to keep themselves in the thermoneutral zone.

Heterothermy

Even though homeotherms regulate their body temperatures around a specific set point, the body temperature of most homeotherms is not completely uniform. Heterothermy describes variations in body temperature along both spatial and temporal scales. For example, animal body temperature is usually warmest at the core but may be much lower in the extremities. The extremities are usually allowed to cool in homeotherms, while the body core temperature can be conserved by warming the blood returning from the extremities through counter-current exchange. In jackrabbits, the ears are allowed to warm above core body temperature, in order to facilitate body heat dissipation by radiation.

A jackrabbit's ears are engorged with blood through vasodilation of capillaries as it promotes heat loss through radiation.

Temporal heterothermy refers to body temperature differences in the same animal over time. Many animals change their insulation values seasonally in order to either conserve or release body heat. Other homeotherms may temporarily use fever in response to pathogen presence. A special case of temporal heterothermy involves animals that not only adjust their body temperature but also adjust their metabolic rates. These adjustments are necessary because energy resources may not be available to fuel the same basal metabolic rates at all times. For example, black-tailed prairie dogs (*Cynomys ludovicianus*) remain active during the winter but periodically undergo shallow bouts of reduced metabolic rate and hypothermia, a state called torpor, when conditions are too harsh to permit exposure. In small bodied-homeotherms, heat loss across their higher surface area relative to body volume is so great that they must undergo daily metabolic adjustments. Daily torpor is a frequent reduction in metabolic rate that conserves limited energy reserves that would otherwise have to be diverted to thermoregulation when the animal could not forage to keep up energy stores.

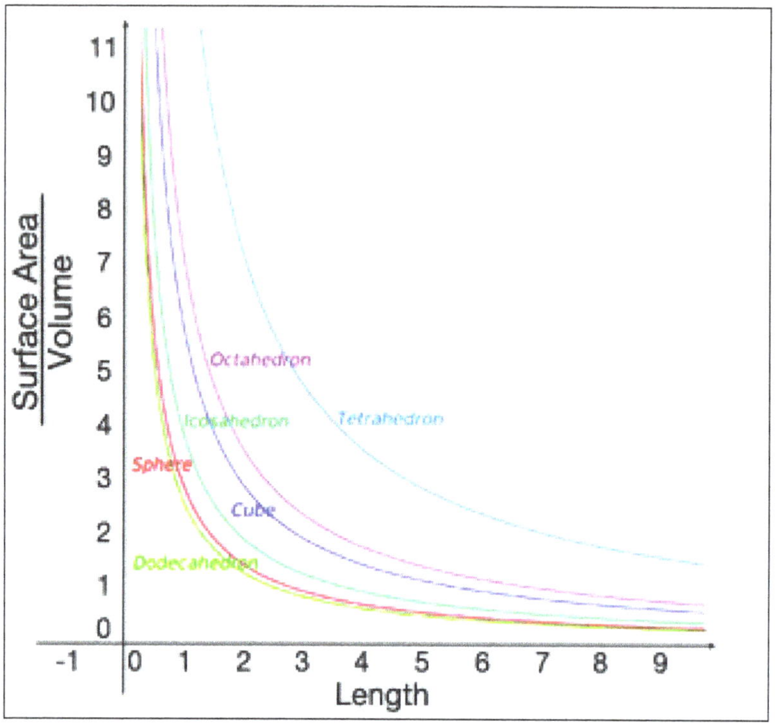

For various shapes, surface area to volume is highest for the smallest length dimensions. The sphere is a useful approximation of an animal body.

Control of Thermoregulation

Both the nervous and endocrine systems control thermoregulatory physiology. Many poikilotherms exhibit periodicity in behavioral thermoregulation; they actively thermoregulate during the day and passively conform during the night described the persistence of a circadian rhythm in selection of thermal microhabitat by sleepy lizards (*Tiliqua rugosa*) in conditions of total darkness. The hormone melatonin (produced by the pineal gland) is implicated in temperature regulation in many ectotherms, while thyroid hormone has a thermogenic effect in endotherms. The "thermostat" for vertebrates resides in the hypothalamus of the brain, which triggers physiological responses to ambient temperatures above and below set points. Indeed, common surgical anesthetics and alcohol can induce hypothermia by disrupting hypothalamic regulation of temperature, especially vasomotor control of peripheral blood vessels.

Life in Extreme Temperatures

When the options of migration and metabolic adjustments are not feasible, resident homeotherms are capable of withstanding extreme temperatures. Reindeer (*Rangifer tarandus*) are notable in remaining active in extremely cold (-23 °C) environments, even having young during the height of the winter. Their thick fur helps with insulation, while regional heterothermy conserves heat in the body core. In addition, their thermoneutral zone extends much further into lower temperatures than in other vertebrates. Indeed, the metabolic rate of a reindeer in winter pelage is lower than that of a reindeer in the summer. The opposite problem is faced by large mammals, such as camels and oryxes that need to withstand extreme heat but be diurnally active. Like reindeer, camels also have thick fur, but this insulation is to prevent ambient heat from the atmosphere entering the body from convection and radiation. Camels and oryxes become hyperthermic (with a body temperature as high as 41 °C) during the heat of the day to reduce the gradient for heat entry into their body.

Tolerance to extreme temperatures is not just a feat for homeotherms. Some poikilotherms are capable of doing this as well. Several species of amphibians and fish are able to withstand freezing temperatures. For example, the wood frog (*Rana sylvatica*) can freeze solid and remain in a state of suspended animation until the spring thaw. Wood frogs and other freezing-tolerant animals utilize cryoprotectants, such as glucose and urea, to dehydrate cells and limit freezing to extracellular spaces.

References

- Animalcell, cells: micro.magnet.fsu.edu, Retrieved 27, May 2020
- Animal-tissue-types, biology: byjus.com, Retrieved 11, August 2020
- Animal-physiology, animal-research, technologies: accela.eu, Retrieved 08, March 2020
- Animal-circulatory-systems, nutrition-transport-and-homeostasis: organismalbio.biosci.gatech.edu, Retrieved 22, January 2020
- Digestion-eunit-23271034: mishicotffa.org, Retrieved 12, July 2020
- Excretion-systems, suny-wmopen-biology: courses.lumenlearning.com, Retrieved 24, February 2020
- Respiratory-system, science: britannica.com, Retrieved 16, June 2020
- EndocrineSystemXV: cals.arizona.edu, Retrieved 05, March 2020
- Locomotion-in-animals-body-structure, cbse: byjus.com, Retrieved 23, August 2020
- Homeostatic-processes-for-thermoregulation-23592046: nature.com, Retrieved 02, April 2020

4

Animal Reproduction

Reproduction is the biological process by which an organism produces an offspring. It is vital for the continuation of life on Earth. This chapter deals with animal reproduction and elaborates the various concepts such as sexual and asexual reproduction, dimorphism and mimicry in an elaborate manner for the reader's benefit.

Reproduction

Reproduction (or procreation, breeding) is the biological process by which new individual organisms – "offspring" – are produced from their "parents". Reproduction is a fundamental feature of all known life; each individual organism exists as the result of reproduction. There are two forms of reproduction: asexual and sexual.

Production of new individuals along a leaf margin of the miracle leaf plant (*Kalanchoe pinnata*). The small plant in front is about 1 cm (0.4 in) tall. The concept of "individual" is obviously stretched by this asexual reproductive process.

In asexual reproduction, an organism can reproduce without the involvement of another organism. Asexual reproduction is not limited to single-celled organisms. The cloning of an organism is a form of asexual reproduction. By asexual reproduction, an organism creates a genetically similar or identical copy of itself. The evolution of sexual reproduction is a major puzzle for biologists. The two-fold cost of sex is that only 50% of organisms reproduce and organisms only pass on 50% of their genes.

Sexual reproduction typically requires the sexual interaction of two specialized organisms, called gametes, which contain half the number of chromosomes of normal cells and are created by meiosis, with typically a male fertilizing a female of the same species to create a fertilized zygote. This produces offspring organisms whose genetic characteristics are derived from those of the two parental organisms.

Asexual

Asexual reproduction is a process by which organisms create genetically similar or identical copies of themselves without the contribution of genetic material from another organism. Bacteria divide asexually via binary fission; viruses take control of host cells to produce more viruses; Hydras (invertebrates of the order *Hydroidea*) and yeasts are able to reproduce by budding. These organisms often do not possess different sexes, and they are capable of "splitting" themselves into two or more copies of themselves. Most plants have the ability to reproduce asexually and the ant species Mycocepurus smithii is thought to reproduce entirely by asexual means.

Some species that are capable of reproducing asexually, like hydra, yeast and jellyfish, may also reproduce sexually. For instance, most plants are capable of vegetative reproduction—reproduction without seeds or spores—but can also reproduce sexually. Likewise, bacteria may exchange genetic information by conjugation.

Other ways of asexual reproduction include parthenogenesis, fragmentation and spore formation that involves only mitosis. Parthenogenesis is the growth and development of embryo or seed without fertilization by a male. Parthenogenesis occurs naturally in some species, including lower plants (where it is called apomixis), invertebrates (e.g. water fleas, aphids, some bees and parasitic wasps), and vertebrates (e.g. some reptiles, fish, and, very rarely, birds and sharks). It is sometimes also used to describe reproduction modes in hermaphroditic species which can self-fertilize.

Sexual

Hoverflies mating in midair flight

Sexual reproduction is a biological process that creates a new organism by combining the genetic material of two organisms in a process that starts with meiosis, a specialized type of cell division. Each of two parent organisms contributes half of the offspring's genetic makeup by creating haploid gametes. Most organisms form two different types of gametes. In these *anisogamous* species, the two sexes are referred to as male (producing sperm or microspores) and female (producing ova or megaspores). In *isogamous species*, the gametes are similar or identical in form (isogametes), but may have separable properties and then may be given other different names. For example, in the green alga, *Chlamydomonas reinhardtii*, there are so-called "plus" and "minus" gametes. A few types of organisms, such as ciliates, *Paramecium aurelia*, have more than two types of "sex", called syngens.

Most animals (including humans) and plants reproduce sexually. Sexually reproducing organisms have different sets of genes for every trait (called alleles). Offspring inherit one allele for each trait from each parent, thereby ensuring that offspring have a combination of the parents' genes. Diploid having two copies of every gene within an organism, it is believed that "the masking of deleterious alleles favors the evolution of a dominant diploid phase in organisms that alternate between haploid and diploid phases" where recombination occurs freely.

Bryophyte reproduces sexually but its commonly seen life forms are all haploid, which produce gametes. The zygotes of the gametes develop into sporangium, which produces haploid spores. The diploid stage is relatively short compared with that of haploid stage, i.e. *haploid dominance*. The advantage of diploid, e.g. heterosis, only takes place in diploid life stage. Bryophyte still maintains the sexual reproduction during its evolution despite the fact that the haploid stage does not benefit from heterosis at all. This may be an example that the sexual reproduction has a bigger advantage by itself, since it allows gene shuffling (hybrid or recombination between multiple loci) among different members of the species, that permits natural selection of the fit over these new hybrids or recombinants that are haploid forms.

Allogamy

Allogamy is the fertilization of an ovum from one individual with the spermatozoa of another.

Autogamy

Self-fertilization, also known as autogamy, occurs in hermaphroditic organisms where the two gametes fused in fertilization come from the same individual, e.g., some foraminiferans, some ciliates. The term "autogamy" is sometimes substituted for autogamous pollination (not necessarily leading to successful fertilization) and describes self-pollination within the same flower, distinguished from geitonogamous pollination, transfer of pollen to a different flower on the same flowering plant, or within a single monoecious Gymnosperm plant. For example, species *Helonias bullata* suffer from low genetic diversity due to self-fertilization.

Mitosis and Meiosis

Mitosis and meiosis are types of cell division. Mitosis occurs in somatic cells, while meiosis occurs in gametes.

Mitosis The resultant number of cells in mitosis is twice the number of original cells. The number of chromosomes in the offspring cells is the same as that of the parent cell.

Meiosis The resultant number of cells is four times the number of original cells. This results in cells with half the number of chromosomes present in the parent cell. A diploid cell duplicates itself, then undergoes two divisions (tetraploid to diploid to haploid), in the process forming four haploid cells. This process occurs in two phases, meiosis I and meiosis II.

Same-sex

In recent decades, developmental biologists have been researching and developing techniques to facilitate same-sex reproduction. The obvious approaches, subject to a growing amount of activity, are female sperm and male eggs, with female sperm closer to being a reality for humans, given that Japanese scientists have already created female sperm for chickens. "However, the ratio of produced W chromosome-bearing (W-bearing) spermatozoa fell substantially below expectations. It is therefore concluded that most of the W-bearing PGC could not differentiate into spermatozoa because of restricted spermatogenesis." In 2004, by altering the function of a few genes involved with imprinting, other Japanese scientists combined two mouse eggs to produce daughter mice.

Strategies

There are a wide range of reproductive strategies employed by different species. Some animals, such as the human and northern gannet, do not reach sexual maturity for many years after birth and even then produce few offspring. Others reproduce quickly; but, under normal circumstances, most offspring do not survive to adulthood. For example, a rabbit (mature after 8 months) can produce 10–30 offspring per year, and a fruit fly (mature after 10–14 days) can produce up to 900 offspring per year. These two main strategies are known as K-selection (few offspring) and r-selection (many offspring). Which strategy is favoured by evolution depends on a variety of circumstances. Animals with few offspring can devote more resources to the nurturing and protection of each individual offspring, thus reducing the need for many offspring. On the other hand, animals with many offspring may devote fewer resources to each individual offspring; for these types of animals it is common for many offspring to die soon after birth, but enough individuals typically survive to maintain the population. Some organisms such as honey bees and fruit flies retain sperm in a process called sperm storage thereby increasing the duration of their fertility.

Other Types

- Polycyclic animals reproduce intermittently throughout their lives.
- Semelparous organisms reproduce only once in their lifetime, such as annual plants (including all grain crops), and certain species of salmon, spider, bamboo and century plant. Often, they die shortly after reproduction. This is often associated with r-strategists.
- Iteroparous organisms produce offspring in successive (e.g. annual or seasonal) cycles, such as perennial plants. Iteroparous animals survive over multiple seasons (or periodic condition changes). This is more associated with K-strategists.

Asexual vs. Sexual Reproduction

Organisms that reproduce through asexual reproduction tend to grow in number exponentially. However, because they rely on mutation for variations in their DNA, all members of the species have similar vulnerabilities. Organisms that reproduce sexually yield a smaller number of offspring, but the large amount of variation in their genes makes them less susceptible to disease.

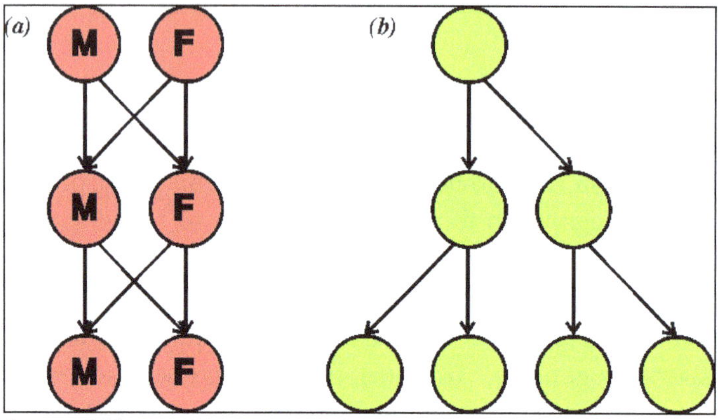

Illustration of the *twofold cost of sexual reproduction*. If each organism were to contribute to the same number of offspring (two), *(a)* the population remains the same size each generation, where the *(b)* asexual population doubles in size each generation.

Many organisms can reproduce sexually as well as asexually. Aphids, slime molds, sea anemones, some species of starfish (by fragmentation), and many plants are examples. When environmental factors are favorable, asexual reproduction is employed to exploit suitable conditions for survival such as an abundant food supply, adequate shelter, favorable climate, disease, optimum pH or a proper mix of other lifestyle requirements. Populations of these organisms increase exponentially via asexual reproductive strategies to take full advantage of the rich supply resources.

When food sources have been depleted, the climate becomes hostile, or individual survival is jeopardized by some other adverse change in living conditions, these organisms switch to sexual forms of reproduction. Sexual reproduction ensures a mixing of the gene pool of the species. The variations found in offspring of sexual reproduction allow some individuals to be better suited for survival and provide a mechanism for selective adaptation to occur. The meiosis stage of the sexual cycle also allows especially effective repair of DNA damages. In addition, sexual reproduction usually results in the formation of a life stage that is able to endure the conditions that threaten the offspring of an asexual parent. Thus, seeds, spores, eggs, pupae, cysts or other "over-wintering" stages of sexual reproduction ensure the survival during unfavorable times and the organism can "wait out" adverse situations until a swing back to suitability occurs.

Life Without

The existence of life without reproduction is the subject of some speculation. The biological study of how the origin of life produced reproducing organisms from non-reproducing elements is called abiogenesis. Whether or not there were several independent abiogenetic events, biologists believe that the last universal ancestor to all present life on Earth lived about 3.5 billion years ago.

Scientists have speculated about the possibility of creating life non-reproductively in the laboratory. Several scientists have succeeded in producing simple viruses from entirely non-living materials. However, viruses are often regarded as not alive. Being nothing more than a bit of RNA or DNA in a protein capsule, they have no metabolism and can only replicate with the assistance of a hijacked cell's metabolic machinery.

The production of a truly living organism (e.g. a simple bacterium) with no ancestors would be a much more complex task, but may well be possible to some degree according to current biological knowledge. A synthetic genome has been transferred into an existing bacterium where it replaced the native DNA, resulting in the artificial production of a new *M. mycoides* organism.

There is some debate within the scientific community over whether this cell can be considered completely synthetic on the grounds that the chemically synthesized genome was an almost 1:1 copy of a naturally occurring genome and, the recipient cell was a naturally occurring bacterium. The Craig Venter Institute maintains the term "synthetic bacterial cell" but they also clarify "...we do not consider this to be "creating life from scratch" but rather we are creating new life out of already existing life using synthetic DNA". Venter plans to patent his experimental cells, stating that "they are pretty clearly human inventions". Its creators suggests that building 'synthetic life' would allow researchers to learn about life by building it, rather than by tearing it apart. They also propose to stretch the boundaries between life and machines until the two overlap to yield "truly programmable organisms". Researchers involved stated that the creation of "true synthetic biochemical life" is relatively close in reach with current technology and cheap compared to the effort needed to place man on the Moon.

Lottery Principle

Sexual reproduction has many drawbacks, since it requires far more energy than asexual reproduction and diverts the organisms from other pursuits, and there is some argument about why so many species use it. George C. Williams used lottery tickets as an analogy in one explanation for the widespread use of sexual reproduction. He argued that asexual reproduction, which produces little or no genetic variety in offspring, was like buying many tickets that all have the same number, limiting the chance of "winning" - that is, producing surviving offspring. Sexual reproduction, he argued, was like purchasing fewer tickets but with a greater variety of numbers and therefore a greater chance of success. The point of this analogy is that since asexual reproduction does not produce genetic variations, there is little ability to quickly adapt to a changing environment. The lottery principle is less accepted these days because of evidence that asexual reproduction is more prevalent in unstable environments, the opposite of what it predicts.

Sexual Reproduction

Sexual reproduction is a form of reproduction where two morphologically distinct types of specialized reproductive cells called gametes fuse together, involving a female's large ovum (or egg) and a male's smaller sperm. Each gamete contains half the number of chromosomes of normal cells. They are created by a specialized type of cell division, which only occurs in eukaryotic cells, known as meiosis. The two gametes fuse during fertilization to produce DNA replication and the creation of a single-celled zygote which includes genetic material from both gametes. In a process called genetic recombination, genetic material (DNA) joins up so that homologous chromosome sequences are aligned with each other, and this is followed by exchange of genetic information. Two rounds of cell division then produce four daughter cells with half the number of chromosomes from each original parent cell, and the same number of chromosomes as both parents, though self-fertilization can occur. For instance, in human reproduction each human cell contains 46 chromosomes,

23 pairs, except gamete cells, which only contain 23 chromosomes, so the child will have 23 chromosomes from each parent genetically recombined into 23 pairs. Cell division initiates the development of a new individual organism in multicellular organisms, including animals and plants, for the vast majority of whom this is the primary method of reproduction. A species is defined as a taxonomic rank. A species is often defined as the largest group of organisms where two hybrids are capable of reproducing fertile offspring, typically using sexual reproduction, although the species problem encompasses a series of difficult related questions that often come up when biologists define the word species.

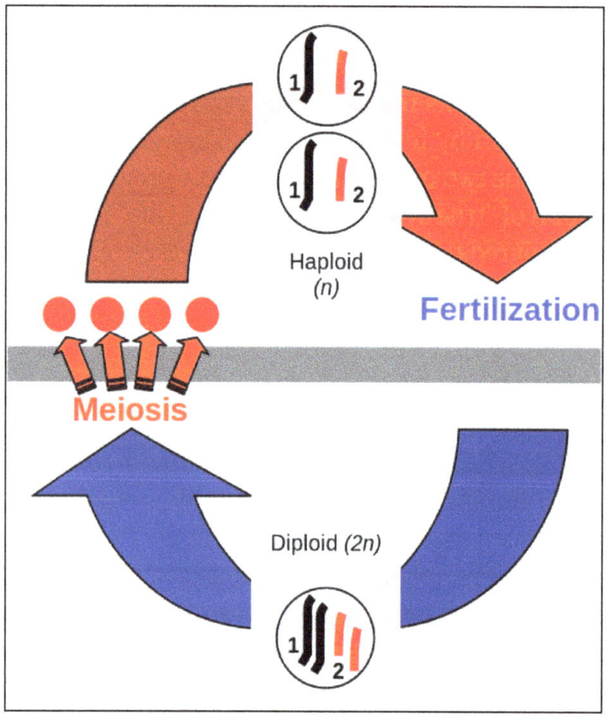

In the first stage of sexual reproduction, "meiosis", the number of chromosomes is reduced from a diploid number (2n) to a haploid number (n). During "fertilization", haploid gametes come together to form a diploid zygote and the original number of chromosomes is restored.

The evolution of sexual reproduction is a major puzzle because asexual reproduction should be able to outcompete it as every young organism created can bear its own young. This implies that an asexual population has an intrinsic capacity to grow more rapidly with each generation. This 50% cost is a fitness disadvantage of sexual reproduction. The two-fold cost of sex includes this cost and the fact that any organism can only pass on 50% of its own genes to its offspring. One definite advantage of sexual reproduction is that it prevents the accumulation of genetic mutations.

Sexual selection is a mode of natural selection in which some individuals out-reproduce others of a population because they are better at securing mates for sexual reproduction. It has been described as "a powerful evolutionary force that does not exist in asexual populations."

Prokaryotes, whose initial cell has additional or transformed genetic material, reproduce through asexual reproduction but may, in lateral gene transfer, display processes such as bacterial conjugation, transformation and transduction, which are similar to sexual reproduction although they do not lead to reproduction.

Evolution

The first fossilized evidence of sexual reproduction in eukaryotes is from the Stenian period, about 1 to 1.2 billion years ago.

Biologists studying evolution propose several explanations for why sexual reproduction developed and why it is maintained. These reasons include fighting the accumulation of deleterious mutations, increasing rate of adaptation to changing environments, dealing with competition, or masking deleterious mutations. All of these ideas about why sexual reproduction has been maintained are generally supported, but ultimately the size of the population determines if sexual reproduction is entirely beneficial. Larger populations appear to respond more quickly to benefits obtained through sexual reproduction than do smaller population sizes.

Maintenance of sexual reproduction has been explained by theories that work at several levels of selection, though some of these models remain controversial.

New models presented in recent years suggest a basic advantage for sexual reproduction in slowly reproducing complex organisms. Sexual reproduction allows these species to exhibit characteristics that depend on the specific environment that they inhabit, and the particular survival strategies that they employ.

Sexual Selection

In order to sexually reproduce, both males and females need to find a mate. Generally in animals mate choice is made by females while males compete to be chosen. This can lead organisms to extreme efforts in order to reproduce, such as combat and display, or produce extreme features caused by a positive feedback known as a Fisherian runaway. Thus sexual reproduction, as a form of natural selection, has an effect on evolution. Sexual dimorphism is where the basic phenotypic traits vary between males and females of the same species. Dimorphism is found in both sex organs and in secondary sex characteristics, body size, physical strength and morphology, biological ornamentation, behavior and other bodily traits. However, sexual selection is only implied over an extended period of time leading to sexual dimorphism.

Sex Ratio

Apart from some eusocial wasps, organisms which reproduce sexually have a 1:1 sex ratio of male and female births. The English statistician and biologist Ronald Fisher outlined why this is so in what has come to be known as Fisher's principle. This essentially says the following:

1. Suppose male births are less common than female.
2. A newborn male then has better mating prospects than a newborn female, and therefore can expect to have more offspring.
3. Therefore parents genetically disposed to produce males tend to have more than average numbers of grandchildren born to them.
4. Therefore the genes for male-producing tendencies spread, and male births become more common.

5. As the 1:1 sex ratio is approached, the advantage associated with producing males dies away.

6. The same reasoning holds if females are substituted for males throughout. Therefore 1:1 is the equilibrium ratio.

Animals

Insects

Australian emperor laying egg, guarded by the male

Insect species make up more than two-thirds of all extant animal species. Most insect species reproduce sexually, though some species are facultatively parthenogenetic. Many insects species have sexual dimorphism, while in others the sexes look nearly identical. Typically they have two sexes with males producing spermatozoa and females ova. The ova develop into eggs that have a covering called the chorion, which forms before internal fertilization. Insects have very diverse mating and reproductive strategies most often resulting in the male depositing spermatophore within the female, which she stores until she is ready for egg fertilization. After fertilization, and the formation of a zygote, and varying degrees of development, in many species the eggs are deposited outside the female; while in others, they develop further within the female and are born live.

Birds

Mammals

There are three extant kinds of mammals: monotremes, placentals and marsupials, all with internal fertilization. In placental mammals, offspring are born as juveniles: complete animals with the sex organs present although not reproductively functional. After several months or years, depending on the species, the sex organs develop further to maturity and the animal becomes sexually mature. Most female mammals are only fertile during certain periods during their estrous cycle, at which point they are ready to mate. Individual male and female mammals meet and carry out copulation. For most mammals, males and females exchange sexual partners throughout their adult lives.

Fish

The vast majority of fish species lay eggs that are then fertilized by the male, some species lay their eggs on a substrate like a rock or on plants, while others scatter their eggs and the eggs are fertilized as they drift or sink in the water column.

Some fish species use internal fertilization and then disperse the developing eggs or give birth to live offspring. Fish that have live-bearing offspring include the guppy and mollies or Poecilia. Fishes that give birth to live young can be ovoviviparous, where the eggs are fertilized within the female and the eggs simply hatch within the female body, or in seahorses, the male carries the developing young within a pouch, and gives birth to live young. Fishes can also be viviparous, where the female supplies nourishment to the internally growing offspring. Some fish are hermaphrodites, where a single fish is both male and female and can produce eggs and sperm. In hermaphroditic fish, some are male and female at the same time while in other fish they are serially hermaphroditic; starting as one sex and changing to the other. In at least one hermaphroditic species, self-fertilization occurs when the eggs and sperm are released together. Internal self-fertilization may occur in some other species. One fish species does not reproduce by sexual reproduction but uses sex to produce offspring; Poecilia formosa is a unisex species that uses a form of parthenogenesis called gynogenesis, where unfertilized eggs develop into embryos that produce female offspring. Poecilia formosa mate with males of other fish species that use internal fertilization, the sperm does not fertilize the eggs but stimulates the growth of the eggs which develops into embryos.

Reptiles

Plants

Animals typically produce gametes directly by meiosis. Male gametes are called sperm, and female gametes are called eggs or ova. In animals, fertilization follows immediately after meiosis. Plants on the other hand have mitosis occurring in spores, which are produced by meiosis. The spores germinate into the gametophyte phase. The gametophytes of different groups of plants vary in size; angiosperms have as few as three cells in pollen, and mosses and other so called primitive plants may have several million cells. Plants have an alternation of generations where the sporophyte phase is succeeded by the gametophyte phase. The sporophyte phase produces spores within the sporangium by meiosis.

Flowering Plants

Flowering plants are the dominant plant form on land and they reproduce either sexually or asexually. Often their most distinguishing feature is their reproductive organs, commonly called flowers. The anther produces pollen grains which contain the male gametophytes (sperm). For pollination to occur, pollen grains must attach to the stigma of the female reproductive structure (carpel), where the female gametophytes (ovules) are located inside the ovary. After the pollen tube grows through the carpel's style, the sex cell nuclei from the pollen grain migrate into the ovule to fertilize the egg cell and endosperm nuclei within the female gametophyte in a process termed double fertilization. The resulting zygote develops into an embryo, while the triploid endosperm (one sperm cell plus two female cells) and female tissues of the ovule give rise to the surrounding tissues in the developing seed. The ovary, which produced the female gametophyte(s), then grows into a fruit, which surrounds the seed(s). Plants may either self-pollinate or cross-pollinate.

Flowers are the sexual organs of flowering plants.

Nonflowering plants like ferns, moss and liverworts use other means of sexual reproduction.

In 2013, flowers dating from the Cretaceous (100 million years before present) were found encased in amber, the oldest evidence of sexual reproduction in a flowering plant. Microscopic images showed tubes growing out of pollen and penetrating the flower's stigma. The pollen was sticky, suggesting it was carried by insects.

Ferns

Ferns mostly produce large diploid sporophytes with rhizomes, roots and leaves; and on fertile leaves called sporangium, spores are produced. The spores are released and germinate to produce short, thin gametophytes that are typically heart shaped, small and green in color. The gametophytes or thallus, produce both motile sperm in the antheridia and egg cells in separate archegonia. After rains or when dew deposits a film of water, the motile sperm are splashed away from the antheridia, which are normally produced on the top side of the thallus, and swim in the film of water to the archegonia where they fertilize the egg. To promote out crossing or cross fertilization the sperm are released before the eggs are receptive of the sperm, making it more likely that the sperm will fertilize the eggs of different thallus. A zygote is formed after fertilization, which grows into a new sporophytic plant. The condition of having separate sporephyte and gametophyte plants is called alternation of generations. Other plants with similar reproductive means include the *Psilotum*, *Lycopodium*, *Selaginella* and *Equisetum*.

Bryophytes

The bryophytes, which include liverworts, hornworts and mosses, reproduce both sexually and vegetatively. They are small plants found growing in moist locations and like ferns, have motile sperm with flagella and need water to facilitate sexual reproduction. These plants start as a haploid spore that grows into the dominate form, which is a multicellular haploid body with leaf-like structures that photosynthesize. Haploid gametes are produced in antherida and archegonia by mitosis. The sperm released from the antherida respond to chemicals released by ripe archegonia and swim to them in a film of water and fertilize the egg cells thus producing a zygote. The zygote divides by mitotic division and grows into a sporophyte that is diploid. The multicellular diploid sporophyte produces structures called spore capsules, which are connected by seta to the archegonia. The spore capsules produce spores by meiosis, when ripe the capsules burst open and the spores are released. Bryophytes show considerable variation in their breeding structures and the above is a basic outline. Also in some species each plant is one sex while other species produce both sexes on the same plant.

Fungi

Fungi are classified by the methods of sexual reproduction they employ. The outcome of sexual reproduction most often is the production of resting spores that are used to survive inclement times and to spread. There are typically three phases in the sexual reproduction of fungi: plasmogamy, karyogamy and meiosis.

Bacteria and Archaea

Three distinct processes in prokaryotes are regarded as similar to eukaryotic sex: bacterial transformation, which involves the incorporation of foreign DNA into the bacterial chromosome; bacterial conjugation, which is a transfer of plasmid DNA between bacteria, but the plasmids are rarely incorporated into the bacterial chromosome; and gene transfer and genetic exchange in archaea.

Bacterial transformation involves the recombination of genetic material and its function is mainly associated with DNA repair. Bacterial transformation is a complex process encoded by numerous bacterial genes, and is a bacterial adaptation for DNA transfer. This process occurs naturally in at least 40 bacterial species. For a bacterium to bind, take up, and recombine exogenous DNA into its chromosome, it must enter a special physiological state referred to as competence. Sexual reproduction in early single-celled eukaryotes may have evolved from bacterial transformation, or from a similar process in archaea.

On the other hand, bacterial conjugation is a type of direct transfer of DNA between two bacteria through an external appendage called the conjugation pilus. Bacterial conjugation is controlled by plasmid genes that are adapted for spreading copies of the plasmid between bacteria. The infrequent integration of a plasmid into a host bacterial chromosome, and the subsequent transfer of a part of the host chromosome to another cell do not appear to be bacterial adaptations.

Exposure of hyperthermophilic archaeal Sulfolobus species to DNA damaging conditions induces cellular aggregation accompanied by high frequency genetic marker exchange. Ajon et al. hypothesized that this cellular aggregation enhances species-specific DNA repair by homologous recombination. DNA transfer in Sulfolobus may be an early form of sexual interaction similar to the more well-studied bacterial transformation systems that also involve species-specific DNA transfer leading to homologous recombinational repair of DNA damage.

Asexual Reproduction

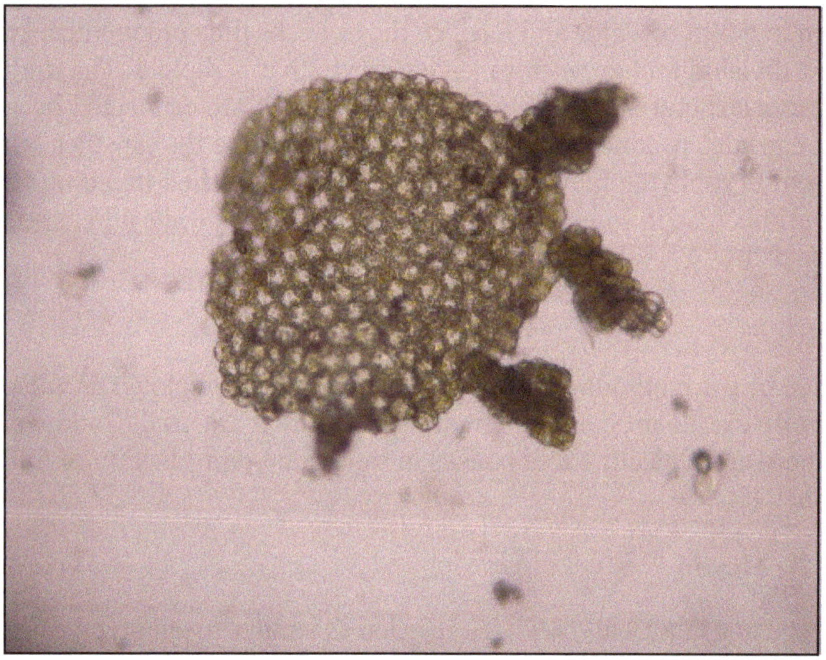

Asexual reproduction in liverworts: a caducous phylloid germinating

Asexual reproduction is a type of reproduction by which offspring arise from a single organism, and inherit the genes of that parent only; it does not involve the fusion of gametes and almost never changes the number of chromosomes. Asexual reproduction is the primary form of reproduction for single-celled organism as the archaea, bacteria, and protists. Many plants and fungi reproduce asexually as well.

While all prokaryotes reproduce asexually (without the formation and fusion of gametes), mechanisms for lateral gene transfer such as conjugation, transformation and transduction are sometimes likened to sexual reproduction (or at least with sex, in the sense of genetic recombination). A complete lack of sexual reproduction is relatively rare among multicellular organisms, particularly animals. It is not entirely understood why the ability to reproduce sexually is so common among them. Current hypotheses suggest that asexual reproduction may have short term benefits when rapid population growth is important or in stable environments, while sexual reproduction offers a net advantage by allowing more rapid generation of genetic diversity, allowing adaptation to changing environments. Developmental constraints may underlie why few animals have relinquished sexual reproduction completely in their life-cycles. Another constraint on switching from sexual to asexual reproduction would be the concomitant loss of meiosis and the protective recombinational repair of DNA damage afforded as one function of meiosis.

Types

Binary Fission

An important form of fission is binary fission. In binary fission, the parent organism is replaced by two daughter organisms, because it literally divides in two. Only prokaryotes (the archaea and the bacteria) reproduce asexually through binary fission. Eukaryotes (such as protists and unicellular fungi) reproduce by mitosis; most of these are also capable of sexual reproduction.

Another type of fission is multiple fission that is advantageous to the plant life cycle. Multiple fission at the cellular level occurs in many protists, e.g. sporozoans and algae. The nucleus of the parent cell divides several times by mitosis, producing several nuclei. The cytoplasm then separates, creating multiple daughter cells.

In apicomplexans, multiple fission, or schizogony, is manifested either as merogony, sporogony or gametogony. Merogony results in merozoites, which are multiple daughter cells, that originate within the same cell membrane, sporogony results in sporozoites, and gametogony results in microgametes.

Budding

Some cells split via budding (for example baker's yeast), resulting in a "mother" and "daughter" cell. The offspring organism is smaller than the parent. Budding is also known on a multicellular level; an animal example is the hydra, which reproduces by budding. The buds grow into fully matured individuals which eventually break away from the parent organism.

Internal budding or Endodyogeny is a process of asexual reproduction, favoured by parasites such as *Toxoplasma gondii*. It involves an unusual process in which two daughter cells are produced inside a mother cell, which is then consumed by the offspring prior to their separation.

Endopolygeny is the division into several organisms at once by internal budding. Also, budding (external or internal) is present in some worm like Taenia or Echinococci; these worm produce cyst and then produce (invaginated or evaginated) protoscolex with budding.

Vegetative Reproduction

Closeup of a Bryophyllum daigremontianum

Vegetative reproduction is a type of asexual reproduction found in plants where new individuals are formed without the production of seeds or spores by meiosis or syngamy. Examples of vegetative reproduction include the formation of miniaturized plants called plantlets on specialized leaves (for example in kalanchoe) and some produce new plants out of rhizomes or stolon (for example in strawberry). Other plants reproduce by forming bulbs or tubers (for example tulip bulbs and dahlia tubers). Some plants produce adventitious shoots and omay form a clonal colony, where all the individuals are clones, and the clones may cover a large area.

Spore

Many multicellular organisms form spores during their biological life cycle in a process called *sporogenesis*. Exceptions are animals and some protists, who undergo *meiosis* immediately followed by fertilization. Plants and many algae on the other hand undergo *sporic meiosis* where meiosis leads to the formation of haploid spores rather than gametes. These spores grow into multicellular individuals (called gametophytes in the case of plants) without a fertilization event. These haploid individuals give rise to gametes through mitosis. Meiosis and gamete formation therefore occur in separate generations or "phases" of the life cycle, referred to as alternation of generations. Since sexual reproduction is often more narrowly defined as the fusion of gametes (fertilization), spore formation in plant sporophytes and algae might be considered a form of asexual reproduction (agamogenesis) despite being the result of meiosis and undergoing a reduction in ploidy. However, both events (spore formation and fertilization) are necessary to complete sexual reproduction in the plant life cycle.

Fungi and some algae can also utilize true asexual spore formation, which involves mitosis giving rise to reproductive cells called mitospores that develop into a new organism after dispersal. This method of reproduction is found for example in conidial fungi and the red algae *Polysiphonia*, and involves sporogenesis without meiosis. Thus the chromosome number of the spore cell is the same as that of the parent producing the spores. However, mitotic sporogenesis is an exception and most spores, such as those of plants, most Basidiomycota, and many algae, are produced by meiosis.

Fragmentation

Fragmentation is a form of asexual reproduction where a new organism grows from a fragment of the parent. Each fragment develops into a mature, fully grown individual. Fragmentation is seen in many organisms such as animals (some planarian and annelid worms,[which?] turbellarians and sea stars), fungi, and plants. Some plants have specialized structures for reproduction via fragmentation, such as *gemma* in liverworts. Most lichens, which are a symbiotic union of a fungus and photosynthetic algae or bacteria, reproduce through fragmentation to ensure that new individuals contain both symbiont. These fragments can take the form of *soredia*, dust-like particles consisting of fungal hyphen wrapped around photobiont cells.

Clonal Fragmentation in multicellular or colonial organisms is a form of asexual reproduction or cloning where an organism is split into fragments. Each of these fragments develop into mature, fully grown individuals that are clones of the original organism. In echinoderms, this method of reproduction is usually known as fissiparity. Researchers claim today that due to many environmental and epigenetic differences, that clones originated in the same ancestor might actually be genetically and epigenetically different.

Agamogenesis

Agamogenesis is any form of reproduction that does not involve a male gamete. Examples are parthenogenesis and apomixis.

Parthenogenesis

Parthenogenesis is a form of agamogenesis in which an unfertilized egg develops into a new individual. Parthenogenesis occurs naturally in many plants, invertebrates (e.g. water fleas, rotifers, aphids, stick insects, some ants, bees and parasitic wasps), and vertebrates (e.g. some reptiles, amphibians, rarely birds). In plants, apomixis may or may not involve parthenogenesis.

Apomixis and Nucellar Embryony

Apomixis in plants is the formation of a new sporophyte without fertilization. It is important in ferns and in flowering plants, but is very rare in other seed plants. In flowering plants, the term "apomixis" is now most often used for agamospermy, the formation of seeds without fertilization, but was once used to include vegetative reproduction. An example of an apomictic plant would be the triploid European dandelion. Apomixis mainly occurs in two forms: In gametophytic apomixis, the embryo arises from an unfertilized egg within a diploid embryo sac that was formed without completing meiosis. In nucellar embryony, the embryo is formed from the diploid nucellus tissue surrounding the embryo sac. Nucellar embryony occurs in some citrus seeds. Male apomixis can occur in rare cases, such as the Saharan Cypress *Cupressus dupreziana*, where the genetic material of the embryo are derived entirely from pollen. The term "apomixis" is also used for asexual reproduction in some animals, notably water-fleas, *Daphnia*.

Alternation between Sexual and Asexual Reproduction

Some species alternate between the sexual and asexual strategies, an ability known as heterogamy, depending on conditions. Alternation is observed in several rotifer species (cyclical parthenogenesis e.g. in Brachionus species) and a few types of insects, such as aphids which will, under certain conditions, produce eggs that have not gone through meiosis, thus cloning themselves. The cape bee *Apis mellifera* subsp. *capensis* can reproduce asexually through a process called thelytoky. A few species of amphibians, reptiles, and birds have a similar ability. For example, the freshwater crustacean *Daphnia* reproduces by parthenogenesis in the spring to rapidly populate ponds, then switches to sexual reproduction as the intensity of competition and predation increases. Another example are monogonont rotifers of the genus *Brachionus*, which reproduce via cyclical parthenogenesis: at low population densities females produce asexually and at higher densities a chemical cue accumulates and induces the transition to sexual reproduction. Many protists and fungi alternate between sexual and asexual reproduction.

For example, the slime mold *Dictyostelium* undergoes binary fission (mitosis) as single-celled amoebae under favorable conditions. However, when conditions turn unfavorable, the cells aggregate and follow one of two different developmental pathways, depending on conditions. In the social pathway, they form a multicellular slug which then forms a fruiting body with asexually generated spores. In the sexual pathway, two cells fuse to form a giant cell that develops into a large cyst. When this macrocyst germinates, it releases hundreds of amoebic cells that are the product of meiotic recombination between the original two cells.

The hyphae of the common mold (*Rhizopus*) are capable of producing both mitotic as well as meiotic spores. Many algae similarly switch between sexual and asexual reproduction. A number of plants use both sexual and asexual means to produce new plants, some species alter their primary modes of reproduction from sexual to asexual under varying environmental conditions.

Inheritance in Sexual Species

For example, in the rotifer *Brachionus calyciflorus* asexual reproduction (obligate parthenogenesis) can be inherited by a recessive allele, which leads to loss of sexual reproduction in homozygous offspring. Inheritance of asexual reproduction by a single recessive locus has also been found in the parasitoid wasp *Lysiphlebus fabarum*.

Examples in Animals

There are examples of parthenogenesis in the hammerhead shark and the blacktip shark. In both cases, the sharks had reached sexual maturity in captivity in the absence of males, and in both cases the offspring were shown to be genetically identical to the mothers. The New Mexico whiptail is another example.

Reptiles use the ZW sex-determination system, which produces either males (with ZZ sex chromosomes) or females (with ZW or WW sex chromosomes). Until 2010, it was thought that the ZW chromosome system used by reptiles was incapable of producing viable WW offspring, but a (ZW) female boa constrictor was discovered to have produced viable female offspring with WW chromosomes. The female boa could have chosen any number of male partners (and had successfully in the past) but on these occasions she reproduced asexually, creating 22 female babies with WW sex-chromosomes.

Polyembryony is a widespread form of asexual reproduction in animals, whereby the fertilized egg or a later stage of embryonic development splits to form genetically identical clones. Within animals, this phenomenon has been best studied in the parasitic Hymenoptera. In the 9-banded armadillos, this process is obligatory and usually gives rise to genetically identical quadruplets. In other mammals, monozygotic twinning has no apparent genetic basis, though its occurrence is common. There are at least 10 million identical human twins and triplets in the world today.

Bdelloid rotifers reproduce exclusively asexually, and all individuals in the class Bdelloidea are females. Asexuality evolved in these animals millions of years ago and has persisted since. There is evidence to suggest that asexual reproduction has allowed the animals to evolve new proteins through the Meselson effect that have allowed them to survive better in periods of dehydration.

Molecular evidence strongly suggest that several species of the stick insect genus *Timema* have used only asexual (parthenogenetic) reproduction for millions of years, the longest period known for any insect.

In the grass thrips genus *Aptinothrips* there have been several transitions to asexuality, likely due to different causes.

Sexual Dimorphism

Female (left) and male (right) common pheasant, illustrating the dramatic difference in both color and size between sexes

Sexual dimorphism is the condition where the two sexes of the same species exhibit different characteristics beyond the differences in their sexual organs. The condition occurs in many animals, insects, birds and some plants. Differences may include secondary sex characteristics, size, color, markings, and may also include behavioral differences. These differences may be subtle or exaggerated, and may be subjected to sexual selection. The opposite of dimorphism is *monomorphism*.

Overview

The peacock, on the right, is courting the peahen, on the left.

Male (bottom) and female mallards. The male mallard has an unmistakable green head.

Orgyia antiqua male and female.

Ornamentation and Coloration

Common and easily identified types of dimorphism are ornamentation and coloration, though not always apparent. A difference in coloration of sexes within a given species is called sexual dichromatism, which is commonly seen in many species of birds and reptiles. Sexual selection leads to the exaggerated dimorphic traits that are used predominantly in competition over mates. The increased fitness resulting from ornamentation offsets its cost to produce or maintain suggesting complex evolutionary implications, but the costs and evolutionary implications vary from species to species.

Exaggerated ornamental traits are used predominantly in the competition over mates, implying sexual selection. Ornaments may be costly to produce or maintain, which has complex evolutionary implications but the costs and implications differ depending on the nature of the ornamentation (such as the colour mechanism involved).

The peafowl constitute conspicuous illustrations of the principle. The ornate plumage of peacocks, as used in the courting display, attracts peahens. At first sight one might mistake peacocks and peahens for completely different species because of the vibrant colours and the sheer size of the male's plumage; the peahen being of a subdued brown coloration. The plumage of the peacock increases its vulnerability to predators because it is a hindrance in flight, and it renders the bird conspicuous in general. Similar examples are manifold, such as in birds of paradise and argus pheasants.

Another example of sexual dichromatism is that of the nestling blue tits. Males are chromatically more yellow than females. It is believed that this is obtained by the ingestion of green lepidopteran larvae, which contain large amounts of the carotenoids lutein and zeaxanthin. This diet also affects the sexually dimorphic colours in the human-invisible UV spectrum. Hence, the male birds, although appearing yellow to humans actually have a violet-tinted plumage that is seen by females. This plumage is thought to be an indicator of male parental abilities. Perhaps this is a good indicator for females because it shows that they are good at obtaining a food supply from which the carotenoid is obtained. There is a positive correlation between the chromas of the tail and breast feathers and body condition. Carotenoids play an important role in immune function for many animals, so carotenoid dependent signals might indicate health.

Frogs constitute another conspicuous illustration of the principle. There are two types of dichromatism for frog species: ontogenetic and dynamic. Ontogenetic frogs are more common and have permanent color changes in males or females. Litoria lesueuri is an example of a dynamic frog that has temporarily color changes in males during breeding season. Hyperolius ocellatus is an ontogenetic frog with dramatic differences in both color and pattern between the sexes. At sexual maturity, the males display a bright green with white dorsolateral lines. In contrast, the females are rusty red to silver with small spots. The bright coloration in the male population serves to attract females and as an aposematic sign to potential predators.

Females often show a preference for exaggerated male secondary sexual characteristics in mate selection. The sexy son hypothesis explains that females prefer more elaborate males and select against males that are dull in color, independent of the species' vision.

Similar sexual dimorphism and mating choice are also observed in many fish species. For example, in guppies males have colorful spots and ornamentations while females are generally grey in color. Female guppies prefer brightly colored males to duller males.

Physiological Differentiation

In redlip blennies, only the male fish develops an organ at the anal-urogenital region that produces antimicrobial substances. During parental care, males rub their anal-urogenital regions over their nests' internal surfaces, thereby protecting their eggs from microbial infections, one of the most common causes for mortality in young fish.

Plants

Most plants are hermaphroditic but approximately 6% have separate males and females (dioecy). Males and females in insect-pollinated species generally look similar to one another because plants provide rewards (e.g. nectar) that encourage pollinators to visit another similar flower, completing pollination. *Catasetum* orchids are one interesting exception to this rule. Male *Catasetum* orchids violently attach pollinia to euglossine bee pollinators. The bees will then avoid other male flowers but may visit the female, which looks different from the males.

Various other dioecious exceptions, such as Loxostylis alata have visibly different genders, with the effect of eliciting the most efficient behaviour from pollinators, who then use the most efficient strategy in visiting each gender of flower instead of searching say, for pollen in a nectar-bearing female flower.

Some plants, such as some species of *Geranium* have what amounts to serial sexual dimorphism. The flowers of such species might for example present their anthers on opening, then shed the exhausted anthers after a day or two and perhaps change their colours as well while the pistil matures; specialist pollinators are very much inclined to concentrate on the exact appearance of the flowers they serve, which saves their time and effort and serves the interests of the plant accordingly. Some such plants go even further and change their appearance again once they have been fertilised, thereby discouraging further visits from pollinators. This is advantageous to both parties because it avoids damage to the developing fruit and avoids wasting the pollinator's effort on unrewarding visits. In effect the strategy ensures that the pollinators can expect a reward every time they visit an appropriately advertising flower.

Females of the aquatic plant *Vallisneria americana* have floating flowers attached by a long flower stalk that are fertilized if they contact one of the thousands of free floating flowers released by a male. Sexual dimorphism is most often associated with wind-pollination in plants due to selection for efficient pollen dispersal in males vs pollen capture in females, e.g. *Leucadendron rubrum*.

Sexual dimorphism in plants can also be dependent on reproductive development. This can be seen in *Cannabis sativa*, a type of hemp, which have higher photosynthesis rates in males while growing but higher rates in females once the plants become sexually mature.

It also should be borne in mind that every sexually reproducing extant species of vascular plant actually has an alternation of generations; the plants we see about us generally are diploid sporophytes, but their offspring really are not the seeds that people commonly recognise as the new generation. The seed actually is the offspring of the haploid generation of microgametophytes (pollen) and megagametophytes (the embryo sacs in the ovules). Each pollen grain accordingly may be seen as a male plant in its own right; it produces a sperm cell and is dramatically different from the female plant, the megagametophyte that produces the female gamete.

Insects

Sexual dimorphism: *Anthocharis cardamines* male is brightly coloured.

Sexual dimorphism: *Anthocharis cardamines* female

Insects display a wide variety of sexual dimorphism between taxa including size, ornamentation and coloration. The female-biased sexual size dimorphism observed in many taxa evolved despite intense male-male competition for mates. In Osmia rufa, for example, the female is larger/broader than males, with males being 8–10 mm in size and females being 10–12 mm in size. The reason for the sexual dimorphism is due to provision size mass, in which females consume more pollen than males. In some species, there evidence of male dimorphism, but it appears to be for the purpose of distinctions of roles. This is seen in the bee species Perdita portalis in which there is a small headed morph, capable of flight, and large headed morph, incapable of flight, for males. *Anthidium manicatum* also displays male-biased sexual dimorphism. The selection for larger size in males rather than females in this species may have resulted due to their aggressive territorial behavior and subsequent differential mating success. Another example is the *L. hemichalceum*, which is a species of sweat bee that shows drastic physical dimorphisms between male offpsring. Not all dimorphism has to have a drastic difference between the sexes. *Andrena agilissima* is a mining bee where the females only have a slightly larger head than the males.

Weaponry leads to increased fitness by increasing success in male-male competition in many insect species. The beetle horns in Onthophagus taurus are enlarged growths of the head or thorax expressed only in the males. Copris ochus also has distinct sexual and male dimorphism in head horns. These structures are impressive because of the exaggerated sizes. There is a direct correlation between male horn lengths and body size and higher access to mates and fitness. In other beetle species, both males and females may have ornamentation such as horns. Generally, insect sexual size dimorphism (SSD) within species increases with body size.

Sexual dimorphism within insects is also displayed by dichromatism. In butterfly genera Bicyclus and Junonia, dimorphic wing patterns evolved due to sex-limited expression, which mediates the intralocus sexual conflict and leads to increased fitness in males. The sexual dichromatic nature of Bicyclus anyana is reflected by female selection on the basis of dorsal UV-reflective eyespot pupils (Robertson & Monteiro, 2005). Naturally selected deviation in protective female coloration is displayed in mimetic butterflies.

Spiders and Sexual Cannibalism

Female (left) and Male (right) *Argiope appensa*, displaying typical sexual differences in spiders, with dramatically smaller males

Size dimorphism shows a correlation with sexual cannibalism, which is prominent in spiders (it is also found in insects such as praying mantises). In the size dimorphic wolf spider, food-limited females cannibalize more frequently. Therefore, there is a high risk of low fitness for males due to pre-copulatory cannibalism, which led to male selection of larger females for two reasons: higher fecundity and lower rates of cannibalism. In addition, female fecundity is positively correlated with female body size and large female body size is selected for, which is seen in the family Araneidae. All Argiope species, including Argiope bruennichi, use this method. Some males evolved ornamentation including binding the female with silk, having proportionally longer legs, modifying the female's web, mating while the female is feeding, or providing a nuptial gift in response to sexual cannibalism. Male body size is not under selection due to cannibalism in all spider species such as Nephila pilipes, but is more prominently selected for in less dimorphic species of spiders, which often selects for larger male size.

Fish

There are cases where males are substantially larger than females. An example is *Lamprologus callipterus*, a type of cichlid fish. In this fish, the males are characterized as being up to 60 times larger than the females. The male's increased size is believed to be advantageous because males collect and defend empty snail shells in each of which a female breeds. Males must be larger and more powerful in order to collect the largest shells. The female's body size must remain small because in order for her to breed, she must lay her eggs inside the empty shells. If she grows too large, she will not fit in the shells and will be unable to breed. Another example is the dragonet, in which males are considerably larger than females and possess longer fins.

The female's small body size is also likely beneficial to her chances of finding an unoccupied shell. Larger shells, although preferred by females, are often limited in availability. Hence, the female is limited to the growth of the size of the shell and may actually change her growth rate according to shell size availability. In other words, the male's ability to collect large shells depends on his size. The larger the male, the larger the shells he is able to collect. This then allows for females to be larger in his brooding nest which makes the difference between the sizes of the sexes less substantial. Male-male competition in this fish species also selects for large size in males. There is aggressive competition by males over territory and access to larger shells. Large males win fights and steal shells from competitors. Sexual dimorphism also occurs in hermaphroditic fish. These species are known as sequential hermaphrodites. In fish, reproductive histories often include the sex-change from female to male where there is a strong connection between growth, the sex of an individual, and the mating system it operates within. In protogynous mating systems where males dominate mating with many females, size plays a significant role in male reproductive success. Males have a propensity to be larger than females of a comparable age but it is unclear whether the size increase is due to a growth spurt at the time of the sexual transition or due to the history of faster growth in sex changing individuals. Larger males are able to stifle the growth of females and control environmental resources.

Social organization plays a large role in the changing of sex by the fish. It is often seen that a fish will change its sex when there is a lack of dominant male within the social hierarchy. The females that change sex are often those who attain and preserve an initial size advantage early in life. In either case, females which change sex to males are larger and often prove to be a good example of dimorphism.

In other cases with fish, males will go through noticeable changes in body size, and females will go through morphological changes that can only be seen inside of the body. For example, in sockeye salmon, males develop larger body size at maturity, including an increase in body depth, hump height, and snout length. Females experience minor changes in snout length, but the most noticeable difference is the huge increase in gonad size, which accounts for about 25% of body mass.

Sexual selection was observed for female ornamentation in Gobiusculus flavescens, known as two-spotted gobies. Traditional hypotheses suggest that male-male competition drives selection. However, selection for ornamentation within this species suggests that showy female traits can be selected through either female-female competition or male mate choice. Since carotenoid-based ornamentation suggests mate quality, female two-spotted guppies that develop colorful orange bellies during the breeding season are considered favorable to males. The males invest heavily in offspring during the incubation, which leads to the sexual preference in colorful females due to higher egg quality.

Amphibians and Reptiles

In amphibians and reptiles, the degree of sexual dimorphism varies widely among taxonomic groups. The sexual dimorphism in amphibians and reptiles may be reflected in any of the following: anatomy; relative length of tail; relative size of head; overall size as in many species of vipers and lizards; coloration as in many amphibians, snakes, and lizards, as well as in some turtles; an ornament as in many newts and lizards; the presence of specific sex-related behaviour is common to many lizards; and vocal qualities which are frequently observed in frogs.

Anolis lizards show prominent size dimorphism with males typically being significantly larger than females. For instance, the average male *Anolis sagrei* was 53.4 mm vs. 40 mm in females. Different sizes of the heads in anoles have been explained by differences in the estrogen pathway. The sexual dimorphism in lizards is generally attributed to the effects of sexual selection, but other mechanisms including ecological divergence and fecundity selection provide alternative explanations. The development of color dimorphism in lizards is induced by hormonal changes at the onset of sexual maturity, as seen in Psamodromus algirus, Sceloporus gadoviae, and S. undulates erythrocheilus.

Birds

Mandarin ducks, male (left) and female (right)

Sexual dimorphism in birds can be manifested in size or plumage differences between the sexes. Sexual size dimorphism varies among taxa with males typically being larger, though this is not always the case i.e. birds of prey, hummingbirds, and some species of flightless birds. Plumage dimorphism, in the form of ornamentation or coloration, also varies, though males are typically the more ornamented or brightly colored sex. Such differences have been attributed to the unequal reproductive contributions of the sexes. This difference produces a stronger female choice since they have more risk in producing offspring. In some species, the male's contribution to reproduction ends at copulation, while in other species the male becomes the main caregiver. Plumage polymorphisms have evolved to reflect these differences and other measures of reproductive fitness, such as body condition or survival. The male phenotype sends signals to females who then choose the 'fittest' available male.

Sexual dimorphism is a product of both genetics and environmental factors. An example of sexual polymorphism determined by environmental conditions exists in the red-backed fairywren. Red-backed fairywren males can be classified into three categories during breeding season: black breeders, brown breeders, and brown auxiliaries. These differences arise in response to the bird's body condition: if they are healthy they will produce more androgens thus becoming black breeders, while less healthy birds produce less androgens and become brown auxiliaries. The reproductive success of the male is thus determined by his success during each year's non-breeding season, causing reproductive success to vary with each year's environmental conditions.

Migratory patterns and behaviors also influence sexual dimorphisms. This aspect also stems back to the size dimorphism in species. It has been shown that the larger males are better at coping with the difficulties of migration and thusly are more successful in reproducing when reaching the breeding destination. When viewing this in an evolutionary standpoint many theories and explanations come to consideration. If these are the result for every migration and breeding season the expected results should be a shift towards a larger male population through sexual selection. Sexual selection is strong when the factor of environmental selection is also introduced. The environmental selection may support a smaller chick size if those chicks were born in an area that allowed them to grow to a larger size, even though under normal conditions they would not be able to reach this optimal size for migration. When the environment gives advantages and disadvantages of this sort, the strength of selection is weakened and the environmental forces are given greater morphological weight. The sexual dimorphism could also produce a change in timing of migration leading to differences in mating success within the bird population. When the dimorphism produces that large of a variation between the sexes and between the members of the sexes multiple evolutionary effects can take place. This timing could even lead to a speciation phenomenon if the variation becomes strongly drastic and favorable towards two different outcomes.

Sexual dimorphism is maintained by the counteracting pressures of natural selection and sexual selection. For example, sexual dimorphism in coloration increases the vulnerability of bird species to predation by European sparrowhawks in Denmark. Presumably, increased sexual dimorphism means males are brighter and more conspicuous, leading to increased predation. Moreover, the production of more exaggerated ornaments in males may come at the cost of suppressed immune function. So long as the reproductive benefits of the trait due to sexual selection are greater than the costs imposed by natural selection, then the trait will propagate throughout the population. Reproductive benefits arise in the form of a larger number of offspring, while natural selection imposes costs in the form of reduced survival. This means that even if the trait causes males to die earlier, the trait is still beneficial so long as males with the trait produce more offspring than males lacking the trait. This balance keeps the dimorphism alive in these species and ensures that the next generation of successful males will also display these traits that are attractive to the females.

Such differences in form and reproductive roles often cause differences in behavior. As previously stated, males and females often have different roles in reproduction. The courtship and mating behavior of males and females are regulated largely by hormones throughout a bird's lifetime. Activational hormones occur during puberty and adulthood and serve to 'activate' certain behaviors when appropriate, such as territoriality during breeding season. Organizational hormones occur only during a critical period early in development, either just before or just after hatching in most birds, and determine patterns of behavior for the rest of the bird's life. Such behavioral differences can cause disproportionate sensitivities to anthropogenic pressures. Females of the whinchat in Switzerland breed in intensely managed grasslands. Earlier harvesting of the grasses during the breeding season lead to more female deaths. Populations of many birds are often male-skewed and when sexual differences in behavior increase this ratio, populations decline at a more rapid rate. Also not all male dimorphic traits are due to hormones like testosterone, instead they are a naturally occurring part of development, for example plumage.

Sexual dimorphism may also influence differences in parental investment during times of food scarcity. For example, in the blue-footed booby, the female chicks grow faster than the males, resulting in booby parents producing the smaller sex, the males, during times of food shortage. This then results in the maximization of parental lifetime reproductive success.

Sexual dimorphism may also only appear during mating season, some species of birds only show dimorphic traits in seasonal variation. The males of these species will molt into a less bright or less exaggerated color during the off breading season. This occurs because the species is more focused on survival than reproduction, causing a shift into a less ornate state.

Consequently, sexual dimorphism has important ramifications for conservation. However, sexual dimorphism is not only found in birds and is thus important to the conservation of many animals. Such differences in form and behavior can lead to sexual segregation, defined as sex differences in space and resource use. Most sexual segregation research has been done on ungulates, but such research extends to bats, kangaroos, and birds. Sex-specific conservation plans have even been suggested for species with pronounced sexual segregation.

The term sesquimorphism (the Latin numeral prefix *sesqui-* means one-and-one-half, so halfway between *mono-* (one) and *di-* (two)) has been proposed for bird species in which "both sexes have basically the same plumage pattern, though the female is clearly distinguishable by reason of her paler or washed-out colour". Examples include Cape sparrow (*Passer melanurus*), rufous sparrow (subspecies *P. motinensis motinensis*), and saxaul sparrow (*P. ammodendri*).

Mammals

Just like in birds, the brains of many mammals, including humans, are significantly different for males and females of the species. Both genes and hormones affect the formation of many animal brains before "birth" (or hatching), and also behaviour of adult individuals. Hormones significantly affect human brain formation, and also brain development at puberty. A 2004 review in *Nature Reviews Neuroscience* observed that "because it is easier to manipulate hormone levels than the expression of sex chromosome genes, the effects of hormones have been studied much more extensively, and are much better understood, than the direct actions in the brain of sex chromosome genes." It concluded that while "the differentiating effects of gonadal secretions seem to be dominant," the existing body of research "support the idea that sex differences in neural expression of X and Y genes significantly contribute to sex differences in brain functions and disease."

Pinnipeds

Marine mammals show some of the greatest sexual size differences of mammals. The mating system of pinnipeds varies from polygyny to serial monogamy. Pinnipeds are known for early differential growth and maternal investment since the only nutrients for newborn pups is the milk provided by the mother. For example, the males are significantly larger than the females at birth in sea lion pups. The pattern of differential investment can be varied principally prenatally and post-natally. *Mirounga leonina*, the southern elephant seal, is one of the most dimorphic mammals.

Sexual dimorphism in elephant seals is associated with the ability of a male to defend territories, which correlates with polygynic behavior. The large sexual size dimorphism is due to sexual selection, but also because females reach reproductive age much earlier than males. In addition the males do not provide parental care for the young and allocate more energy to growth. This is supported by the secondary growth spurt in males during adolescent years.

Primates

Humans

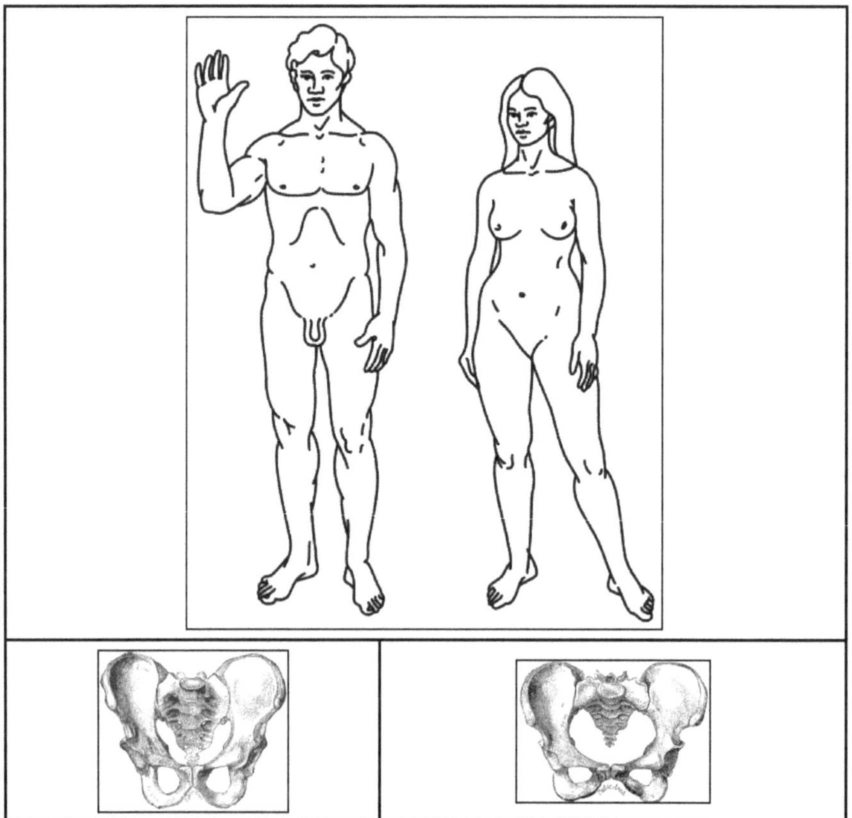

Top: Stylised illustration of humans on the Pioneer plaque, showing both male (left) and female (right).
Bottom: Comparison between male (left) and female (right) pelvises.

In humans, biological sex is determined by five factors present at birth: the presence or absence of a Y chromosome, the type of gonads, the sex hormones, the internal reproductive anatomy (such as the uterus in females), and the external genitalia. Generally, the five factors are either all male or all female. Sexual ambiguity is rare in humans, but wherein such ambiguity does occur, the individual is biologically classified as intersex.

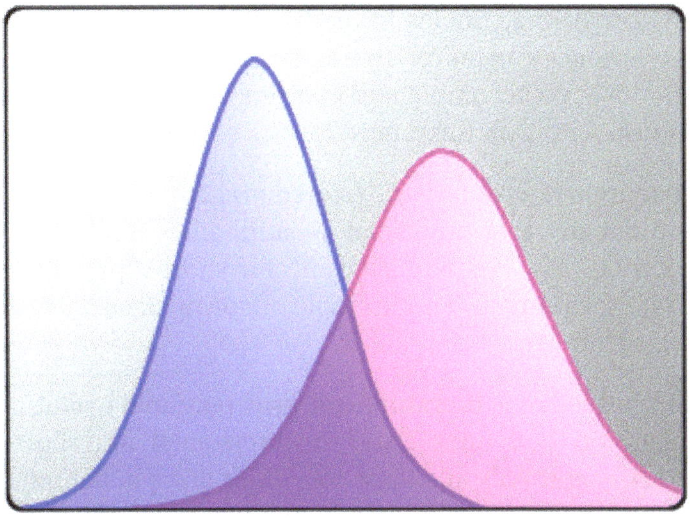

This is a remake of a facial geometric sexual dimorphism diagram from Valenzano, D. R. et al. (2006). The blue bell curve on the left represents the male faces, and the pink bell curve on the right represents the female faces. The purple area in the center represents the overlap of the two bell curves where the feminine male faces cannot be distinguished from the masculine female faces. The bell curves show that the proportion of female faces that are more feminine than the most feminine male faces is much greater than the proportion of male faces that are more masculine than the most masculine female faces.

Sexual dimorphism among humans includes differentiation among gonads, internal genitals, external genitals, breasts, muscle mass, height, the endocrine (hormonal) systems and their physiological and behavioral effects. Human sexual differentiation is effected primarily at the gene level, by the presence or absence of a Y-chromosome, which encodes biochemical modifiers for sexual development in males. According to Clark Spencer Larsen, modern day *Homo sapiens* show a range of sexual dimorphism, with average body mass difference between the sexes being roughly equal to 15% . Ever since Charles Darwin's *The Descent of Man and Selection in Relation to Sex* in 1871 was published, there's been controversy regarding the social, cultural, and political significance of human sexual dimorphism.

The average basal metabolic rate is about 6 percent higher in adolescent males than females and increases to about 10 percent higher after puberty. Females tend to convert more food into fat, while males convert more into muscle and expendable circulating energy reserves. Aggregated data of absolute strength indicates that females have, on average, 40-60% the upper body strength of males, and 70-75% the lower body strength. The difference in strength relative to body mass is less pronounced in trained individuals. In Olympic weightlifting, male records vary from 5.5× body mass in the lowest weight category to 4.2× in the highest weight category, while female records vary from 4.4× to 3.8×, a weight adjusted difference of only 10-20%, and an absolute difference of about 30% (i.e. 472 kg vs 333 kg for unlimited weight classes). A study, carried about by analyzing annual world rankings from 1980–1996, found that males' running times were, on average, 11% faster than females'.

Females are taller, on average, than males in early adolescence, but males, on average, surpass them in height in later adolescence and adulthood. In the United States, adult males are, on average, 9% taller and 16.5% heavier than adult females.

Males typically have larger tracheae and branching bronchi, with about 30 percent greater lung volume per body mass. On average, males have larger hearts, 10 percent higher red blood cell count, higher hemoglobin, hence greater oxygen-carrying capacity. They also have higher circulating clotting factors (vitamin K, prothrombin and platelets). These differences lead to faster healing of wounds and higher peripheral pain tolerance.

Females typically have more white blood cells (stored and circulating), more granulocytes and B and T lymphocytes. Additionally, they produce more antibodies at a faster rate than males. Hence they develop fewer infectious diseases and succumb for shorter periods. Ethologists argue that females, interacting with other females and multiple offspring in social groups, have experienced such traits as a selective advantage.

Considerable discussion in academic literature concerns potential evolutionary advantages associated with sexual competition (both intrasexual and intersexual) and short- and long-term sexual strategies. According to Daly and Wilson, "The sexes differ more in human beings than in monogamous mammals, but much less than in extremely polygamous mammals." One proposed explanation is that human sexuality has developed more in common with its close relative the bonobo, who have similar sexual dimorphism and which are polygynandrous and use recreational sex to reinforce social bonds and reduce aggression.

In the human brain, a difference between sexes was observed in the transcription of the PCDH11X-/Y gene pair unique to *Homo sapiens*. Sexual differentiation in the human brain from the default female state is triggered by testosterone from the fetal testis. Testosterone is converted to estrogen in the brain through the action of the enzyme aromatase. Testosterone acts on many brain areas, including the SDN-POA, to create the masculinized brain pattern. Female brains may be shielded from the masculinizing effects of estrogen through the action of sex-hormone binding globulin and a-fetoprotein.

The relationship between sex differences in the brain and human behavior is a subject of controversy in psychology and society at large. Many females tend to have a higher ratio of gray matter in the left hemisphere of the brain in comparison to males. Males on average have larger brains than females, however when adjusted for total brain volume the gray matter differences between sexes is almost nonexistent. Thus, the percentage of gray matter appears to be more related to brain size than it is to sex. Differences in brain physiology between sexes do not necessarily relate to differences in intellect. Haier *et al.* found in a 2004 study that "men and women apparently achieve similar IQ results with different brain regions, suggesting that there is no singular underlying neuroanatomical structure to general intelligence and that different types of brain designs may manifest equivalent intellectual performance".

Cells

Phenotypic differences between sexes are evident even in cultured cells from tissues. For example, female muscle-derived stem cells have a better muscle regeneration efficiency than male ones. There are reports of several metabolic differences between male and female cells and they also respond to stress differently.

Larger Females are Reproductively Advantageous

In theory, larger females are favored by competition for mates, especially in polygamous species. Larger females offer an advantage in fertility, since the physiological demands of reproduction are limiting in females. Hence there is a theoretical expectation that females tend to be larger in species that are monogamous. Females are larger in many species of insects, many spiders, many fish, many reptiles, owls, birds of prey and certain mammals such as the spotted hyena, and baleen whales such as blue whale. As an example, in some species, females are sedentary and sparsely distributed, and so males must search for them. Fritz Vollrath and Geoff Parker argue that this difference in behaviour leads to radically different selection pressures on the two sexes, evidently favouring smaller males. Cases where the male is larger than the female have been studied as well, and require alternative explanations.

One example of this type of sexual size dimorphism is the bat *Myotis nigricans*, where females are substantially larger than males in terms of body weight, skull measurement, and forearm length. The interaction between the sexes and the energy needed to produce viable offspring make it favorable for females to be larger in this species. Females bear the energetic cost of producing eggs, which is much greater than the cost of making sperm by the males. The fecundity advantage hypothesis states that a larger female is able to produce more offspring and give them more favorable conditions to ensure their survival; this is true for most ectotherms. A larger female can provide parental care for a longer time while the offspring matures. The gestation and lactation periods are fairly long in *M. nigricans*, the females suckling their offspring until they reach nearly adult size. They would not be able to fly and catch prey if they did not compensate for the additional mass of the offspring during this time. Smaller male size may be an adaptation to increase maneuverability and agility, allowing males to compete better with females for food and other resources.

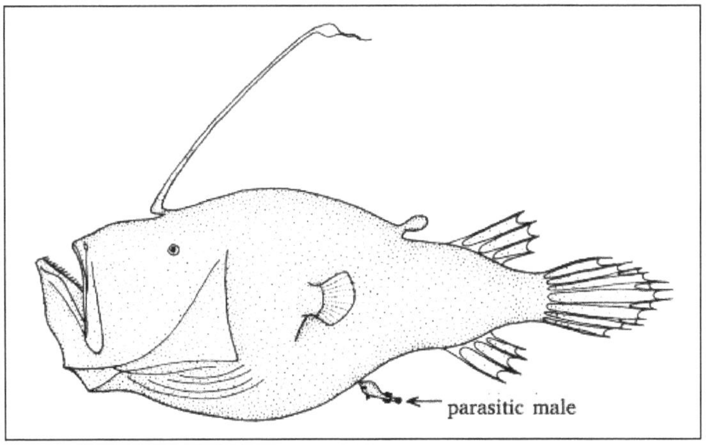

Female triplewart seadevil, an anglerfish, with male attached near vent (arrow)

Some species of anglerfish also display extreme sexual dimorphism. Females are more typical in appearance to other fish, whereas the males are tiny rudimentary creatures with stunted digestive systems. A male must find a female and fuse with her: he then lives parasitically, becoming little more than a sperm-producing body in what amounts to an effectively hermaphrodite composite organism. A similar situation is found in the Zeus water bug *Phoreticovelia disparata* where the female has a glandular area on her back that can serve to feed a male, which clings to her (note that although males can survive away from females, they generally are not free-living).

Some plant species also exhibit dimorphism in which the females are significantly larger than the males, such as in the moss *Dicranum* and the liverwort *Sphaerocarpos*. There is some evidence that, in these genera, the dimorphism may be tied to a sex chromosome, or to chemical signalling from females.

Another complicated example of sexual dimorphism is in the *Vespula squamosa*, or southern yellowjacket. In this wasp species, the female workers are the smallest, the male workers are slightly larger, and the female queens are significantly larger than her female worker and male counterparts.

Evolution

Sexual dimorphism by size is evident in some extinct species such as the velociraptor. In the case of velociraptors the sexual size dimorphism may have been caused by two factors: male competition for hunting ground to attract mates, and/or female competition for nesting locations and mates, males being a scarce breeding resource.

In 1871, Charles Darwin advanced the theory of sexual selection, which related sexual dimorphism with sexual selection.

It has been proposed that the earliest sexual dimorphism is the size differentiation of sperm and eggs (anisogamy), but the evolutionary significance of sexual dimorphism is more complex than that would suggest. Anisogamy and the usually large number of small male gametes relative to the larger female gametes usually lies in the development of strong sperm competition, because small sperm enable organisms to produce a large number of sperm, and make males (or male function of hermaphrodites) more redundant. This intensifies male competition for mates and promotes the evolution of other sexual dimorphim in many species, especially in vertebrates including mammals. However, in some species, the females can be larger than males, irrespective of gametes, and in some species females (usually of species in which males invest a lot in rearing offspring and thus no longer considered as so redundant) compete for mates in ways more usually associated with males.

In many non-monogamous species, the benefit to a male's reproductive fitness of mating with multiple females is large, whereas the benefit to a female's reproductive fitness of mating with multiple males is small or nonexistent. In these species, there is a selection pressure for whatever traits enable a male to have more matings. The male may therefore come to have different traits from the female.

Male (left), offspring, and female (right) Sumatran orangutans.

These traits could be ones that allow him to fight off other males for control of territory or a harem, such as large size or weapons; or they could be traits that females, for whatever reason, prefer in mates. Male-male competition poses no deep theoretical questions but mate choice does.

Females may choose males that appear strong and healthy, thus likely to possess "good alleles" and give rise to healthy offspring. In some species, however, females seem to choose males with traits that do not improve offspring survival rates, and even traits that reduce it (potentially leading to traits like the peacock's tail). Two hypotheses for explaining this fact are the sexy son hypothesis and the handicap principle.

The sexy son hypothesis states that females may initially choose a trait because it improves the survival of their young, but once this preference has become widespread, females must continue to choose the trait, even if it becomes harmful. Those that do not will have sons that are unattractive to most females (since the preference is widespread) and so receive few matings.

The handicap principle states that a male who survives despite possessing some sort of handicap thus proves that the rest of his genes are "good alleles". If males with "bad alleles" could not survive the handicap, females may evolve to choose males with this sort of handicap; the trait is acting as a hard-to-fake signal of fitness.

Sexual Mimicry

Sexual mimicry occurs when one sex mimics the opposite sex in its behavior, appearance, or chemical signalling. It is more commonly seen within invertebrate species, although sexual mimicry is also seen among vertebrates such as spotted hyenas. Sexual mimicry is commonly used as a mating strategy to gain access to a mate, a defense mechanism to avoid more dominant individuals, or a survival strategy. It can also be a physical characteristic that establishes an individual's place in society. Sexual mimicry is employed differently across species and it is part of their strategy for survival and reproduction. Examples of sexual mimicry in animals include the spotted hyena, certain types of fish, passerine birds and some species of insect among others. These are cases of intraspecific sexual mimicry, but interspecific sexual mimicry can also occur in some plant species, especially orchids. In plants employing sexual mimicry, flowers mimic mating signals of their pollinator insects. These insects are attracted and pollinate the flowers through pseudocopulations or other sexual behaviors performed on the flower.

Social Systems

Sexual mimicry can influence the species' social system. The most common example is the spotted hyenas, *Crocuta crocuta*. Female hyenas resemble male hyenas in their sexual anatomy: the females have peniform clitorises, resembling a penis, and false scrotal sacs. These characteristics, as well as high androgen levels in their blood, make for aggressive females, which results in their dominance over males; the female with the lowest rank is more dominant than the highest-ranking male. Within the female population in each clan, there are different ranks: the dominant females, who reproduce at an earlier age and get more access to food, and the non-dominant females. Their dominance is hierarchical and is passed from mother to daughter. By contrast, male spotted hyenas gain their social status with the length of their stay in the clan; it does not involve aggressive contests. The males leave their clan between the ages of two and six and join a different clan where they gain status with age. Males also foster amicable relationships with the females to stabilize their position in the social hierarchy.

Hyena greeting ceremony

Because females are the dominant sex among spotted hyenas, they are the most respected. Subordinate female hyenas initiate a 'greeting' with dominant female hyenas as a sign of respect and are forced to do so if they refuse. This greeting used by hyenas reflects the asymmetry of their ranking; the animal being greeted (the subordinate individual) extends its hind legs and the individual doing the greeting (the dominant hyena) licks or sniffs the erect peniform clitoris. By lifting its hind legs, the hyena being greeted (the subordinate hyena) exposes its most vulnerable body part to the other individual, an act that reflects inferiority. As well, when its hind legs are lifted, a scent can be identified by the other individual. Subordinate hyenas expose their scent more often than high-ranking hyenas. This greeting, however, is not commonly seen between males and adult females; when it does occur, it is restricted to males of median or higher rank greeting dominant females.

Mating Systems

In the spotted hyenas, the only way for the males to mate with the females is if they have the female's full cooperation because of the female's peniform clitoris. An increase in the male's status gave them more access to dominant females in the clan. Female dominant hyenas do not mate with multiple males, possibly due to the cost of cleaning their genitalia, which hyenas are seen doing after copulation. Because they will get access to the most dominant and better fit males, they do not need to copulate with multiple males to produce offspring of higher fitness. Non-dominant females are observed copulating more often with lower-ranking males. It is costly for female hyenas to give birth through their long peniform clitoris. The umbilical cord is 12–18 cm long, while the journey from the uterus to the clitoris end is 40 cm. The umbilical cord often breaks before the cub emerges, leading to death by anoxia for many young. This journey is not only harmful for the cubs, but also for the mother. The tissue of the clitoris will sometimes rip open when giving birth for the first time which can be fatal to the mother.

Female spotted hyenas are the choosy sex because they invest in parental care as well as being the dominant sex in the clan. However, males are likely to still have a preference for a particular female as it is seen in other animals; high-ranking females begin breeding at a younger age and their offspring are more likely to survive to adulthood than the offspring of low-ranking females. Males associate more closely with females that are fertile, a state most likely noticed through olfactory cues. While middle/high-ranking males associate with high-ranking females, low-ranking males associate equally with high and low-ranking females. Associating with low-ranking females may be due to low-ranking males failing to recognize the reproductive success of high-ranking females or using a different type of reproductive strategy. Males tend to spend lot of time with the female they mate with before conception to avoid other males coming in close contact with her.

Sexual mimicry is also used as a mate-guarding strategy by some species. Mate-guarding is a process in which a member of a species prevents another member of the same species from mating with their partner. Mate-guarding is seen in *Cotesia rubecula*, a parasitic wasp from the Braconidae family whose mating system is polygynous. Males are attracted to females through pheromones and they induce females to mate through vibrations, to which the female responds by assuming a specific position. When a male who has copulated with a female sees another male trying to court her, he will often adopt the female receptive position. Post-copulatory female mimicry by the male offers an advantage by acting as a mate-guarding mechanism. If a second male arrives soon enough after the female copulates with the first male, the second male may be able to induce a second copulation which will compete with the first one. However, if the first male who copulated with her mimics the female, it distracts the second male long enough that the female becomes unreceptive.

Sneaky Copulation

Salaria pavo female

Salaria pavo male

Sneaky copulation is a strategy used by many aquatic organisms who portray sexual mimicry. Several studies have found that small male fish will look and behave like the female of their species in order to gain access to female territory and copulate with them. In the Blenniidae family, the *Salaria pavo*, female bleniid fish will show a specific colour pattern and movement when they want to approach a male and copulate with him. The male guards a territory, and when the female lays her eggs, the parental male protects that territory until the eggs hatch. A second type of males, the sneaker males, is parasitic and resembles the female bleniid fish in their small size, colour, and movement patterns. This allows them to intrude into the nest guarded by the parental males. Sneaker males approach the nests with the same colour patterns and movements that the females hold. Most cases of sneaker males are seen when there is a female already inside the nest although sometimes the sneaker fish enters the nest alongside a female. This species of fish releases the sperm before the female releases her eggs into the water making it possible for the sneaker fish to fertilize an egg, even if the female is not present in the nest.

Sepia apama

In the Sepiina family, *Sepia apama*, also known as cuttlefish, have some males that are large and able to guard a female's nest while other males are small and resemble females in order to sneak in copulations. In the giant cuttlefish, the male courts the female and transfers its sperm to a pouch below the female's beak. During this process, the female displays a body pattern of black splotches on a white background. Once the eggs are laid, the male guards the nest from any possible suitors and opponents. A 'second female' is sometimes seen during male-female interaction in close proximity to the couple. This female-looking cuttlefish has the same black blotches as a real female. If the male leaves to fight other males, this individual approaches the female and copulates with her, usually with success. However, in the absence of rivals, these 'mimicking female' males display the phenotype of a mature male.

Sexual Mimicry against Aggression

A similar phenomenon to the sneaker fish males is observed in the dark-edged splitfin, *Girardinichthys multiradiatus*. The juveniles resemble the pregnant females in the species by having a dark spot near the vent. In this case, however, the mimicking males have the capability to resemble the females or become a morphologically mature male throughout most of their adult life. This dark spot allows the female-looking males to escape aggression from more dominant males, as well as reducing the chance of having a female nearby flee due to persisting courting males. The mature males do not attack the subordinate fish and the subordinate fish decides when to initiate the fights, which gives it an advantage as the mature male is not expecting this. The dark spot also permits access of subordinate males to females, a characteristic that is advantageous because females' eggs can only be fertilized during a five-day fertilizing window.

Male *Ficedula hypoleuca*

Female/juvenile *Ficedula hypoleuca*

Sexual mimicry to avoid aggression is also seen in birds. In some bird species, males have a female-like plumage colour during their second year of life (SY males). These SY males are sexually mature and able to breed, but their morphology differs greatly from the older, after second year (ASY) males. Various studies have looked into this delayed plumage maturation (DPM) and found that the DPM in SY males reduces aggression from ASY males. Female mimicry in birds was first found in European-pied flycatcher, *Ficedula hypoleuca*. When a dull-coloured male is in the area, mature males reduce their aggressiveness and behave as if the intruder is a female. The dull plumage is seen mostly in younger males, likely due to being born later in the previous spring. The resemblance to females benefit these young males when trying to occupy a territory with many males already present because the young males can gain information and access to a territory that would not be accessible to them otherwise.

There is a big cost to not looking like a male when it comes to defending a territory or attracting a mate. Females show aggression against dull-coloured males, making it harder for them to mate. However, DPM has some benefits: as mentioned above, it reduces aggression from older males. As well, these female-looking birds are able to get access to territories, mates, and food that may be not be available to them otherwise. Another benefit is that DPM provides SY birds with a longer lifespan; because they do not have to compete with other males, their mortality rate is lower. This advantage, however, only benefits individuals of species that have a longer potential lifespan and, therefore, DPM would not benefit a short-lived species. This is known as the breeding threshold hypothesis, and states that SY males should only delay breeding if there is a large mortality difference between the SY males who attempt to breed and the ones who do not.

Platysaurus broadleyi

Most studies addressed DPM as a type of sexual mimicry, which is done through deception: male ASY birds should not be able to tell females or SY males apart. However, Muheter et al. (1997) found that territorial males perceive the dull-coloured males as males but they show less aggression because their dull-coloured plumage promotes low competitive ability. They referred to this as honest signalling and not sexual mimicry.

Another example of sexual mimicry occurs in Broadley's Flat Lizard, *Platysaurus broadleyi*, where some males mimic females. Flat lizard males tend to be territorial and aggressive towards other males. Therefore, it is beneficial for some males to mimic females in order to avoid aggressive encounters and move freely through the male's territory, looking for mates. There are two types of males in this population; she-males, who mimic females, and he-males, who look like males. The she-males can visually fool the he-males into believing that they are female due to their female morphology. However, the she-males cannot fool the he-males through scent, as he-males can detect the difference. Therefore, the most successful she-males are those who avoid close contact with other males, thereby reducing the chances of detection through chemical signals.

Molecular Control over Sexual Mimicry

Female hyenas' sexual mimicry to males is part of their anatomy and it is thought to have evolved through high androgen levels. While female ancestors were smaller than males, selection must have acted upon androgen levels and female body size to increase both, leading to further selection and larger females than males. The high androgen levels are not present in the female ovaries, as it was once thought; the stromal tissue in the ovaries contains lower testosterone levels than the males' testes. However, females' androgen levels in the blood are as high as the ones found in the male, having the effect of morphologically male-looking females.

Ruffs can also show sexual mimicry through a combination of genetics and hormones. In a population of ruffs, *Philomachus pugnax*, there are three types of male morphs: independent males and satellite males, both of which are reproductive competitors, and faeder ruffs that resemble females in their plumage. The first two morphs are controlled by a dominant allele at a single autosomal locus, while the third morph is likely to have come from a combination of a third allele and a lack of testosterone. When testosterone is administered to reeves (female ruffs), male courtship behaviour and male feather colouration are expressed in the reeves. Testosterone, in this case, expresses sex-limited characteristics by acting on the single autosomal gene. Similarly, while it has not yet been tested, it is likely that the lack of testosterone is the cause for the faeder ruffs' similarity to females.

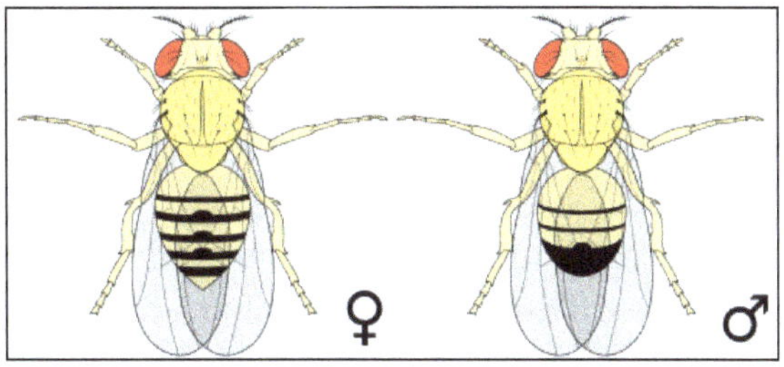

Biology Illustration Animals Insects *Drosophila melanogaster*

A different example is seen in mature female fruit flies, *Drosophila melanogaster*, who are very attractive but their level of attractiveness decreases by half or more after three minutes of mating. Males release a compound, 7-tricosene, into the female during courtship that lowers female attractiveness. However, the researcher found that the females release this compound as well, six hours after mating. This compound lowers the female's levels of attractiveness both times, when the male is courting her and during mating. This way, the female mimics the male and with this compound, she lowers her levels of attractiveness .

Genetic Control Over Sexual Mimicry

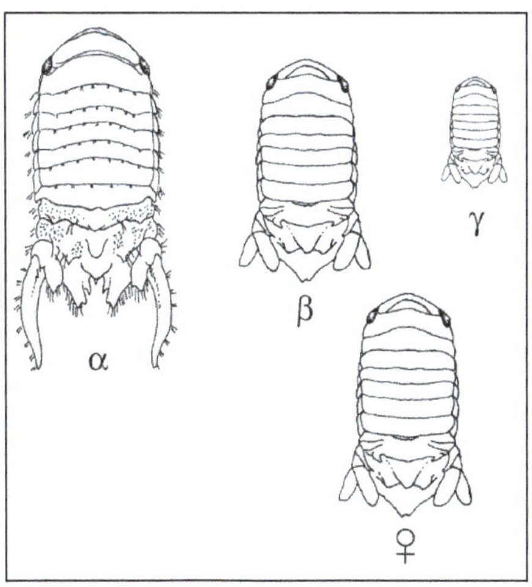

Paracerceis sculpta (Marine isopod)

Some organisms' sexual mimicry is genetically determined by specific alleles. Unlike sexual mimicry that arises due to molecular compounds or hormones and can sometimes be induced through these molecules, this sexual mimicry arises from the organism's genetic material. Besides the female hyenas' sexual anatomy, which is part of their genetics, some other organisms have only some males/females in their population who look like the opposite sex and this is determined by specific alleles.

In the marine isopod population, *Paracerceis sculpta*, there are three different male morphologies: the alpha male is the largest morph, it matures last, and it is the one who gets privileged access to the females. The beta male is of intermediate size, and it mimics the female to get access to females. Last, the gamma male is the smallest morph and it invades harems, where females go to mate with alpha males, for mating opportunities. This morphology is associated with a single autosomal gene and three different alleles. Beta is the most dominant allele, followed by gamma, which is followed by alpha. Selection on these alleles acts according to the Hardy-Weinberg equilibrium and mating success is equivalent among all three morphs.

The alpha males, who are homozygous for the alpha allele, mate with many females in a harem. The females prefer to aggregate with other females in the harem, which gives the alpha male a bigger selection of mating partners. Shuster (1992) looked at the behaviour and relationship of each morph with respect to the harem and found that beta and gamma males could locate harems that have sexually receptive females. They were also able to differentiate between a harem with a sexually receptive female, i.e. one that is able to mate, and a non-sexually receptive female, i.e. one that has already deposited the embryo into her pouch and can no longer mate. While it is still unclear how the beta males do this or how their mating strategies work, they are not harassed by alpha males due to their mimicry of females: the beta males can attract other females into the harem since females like to go where other females are, and this provides the alpha males with more mates.

Another order of organisms whose sexual mimicry is influenced by their DNA is the Odonata, carnivorous insects known as dragonflies and damselflies. In these species, it is the female who sometimes mimics the male. Within a species, groups of females will differ in colour: one group mimics the males' colour and they are known as androchromes. Other groups will have their own female colouration and they are known as gynochromes. In *Ischnura elegans*, androchromes comprise 6-30% of the female population and their colour is usually blue, like the males; in some populations, androchromes are larger in size than gynochromes. This polymorphism is controlled by an autosomal allele and some studies have looked at the reason for the polymorphism's maintenance.

Aeshna affinis Male

Aeshna affinis Female Androchrome

The most likely theory for the maintenance of the polymorphism in Odonata is the density dependence theory that states that at a high male density, the androchromes are not bothered by the males and their existence is not threatened by male harassment. This hypothesis also assumes that males cannot distinguish between androchromes and other males. This advantage, however, is counteracted with the fact that they will not get a lot of mating opportunities (if any) and their reproduction is limited. This theory is the most likely explanation for the maintenance of polymorphism, since studies have shown that there is an advantage for androchromes in high male-density populations.

Self-control over Sexual Mimicry

While, as seen before, most organisms which portray sexual mimicry are born with this morphology/behaviour, this is not always the case. The giant cuttlefish, *Sepia apama*, mentioned above in the section "sneaky copulations", is born with the capacity to choose whether to change its morphology to look like a female or a mature male. When no competition is seen nearby, the cuttlefish will look like a mature male and mate with the female. However, when a mature male and a female are copulating, the giant cuttlefish will resemble a female and stay at a close distance of the couple, hoping for a chance to mate with the female if the mature male leaves to fight other males. Another example of an organism that has the capability to remain small and look like a female, or become a morphologically mature male, is the dark-edged splitfin, *Girardinichthys multiradiatus*. The purpose for their female mimicry was seen before, in the "sexual mimicry against aggression" section where the female-looking males will escape aggression from dominant males and avoid females fleeing their company due to persisting courting males.

Interspecific Deceptive Mimicry

Interspecific sexual mimicry can also occur in some plant species. The most common example of this is known as sexually deceptive pollination and is found among some orchids. The orchid mimics its pollinator's females, usually hymenopterans such as wasps and bees, attracting the males to the flower. Orchid flowers mimic the sex pheromones and to some degree the visual appearance of the female insect of its pollinator species. The primacy of olfactory over visual cues has been demonstrated in many cases, such as in the European orchid genus Ophrys as well as many Australian sexually deceptive orchids.

Bee Orchid

In few other cases, such as the South African daisy *Gorteria diffusa*, visual signals seem to be of primary importance. Visual signals also enhance the attractiveness of the flowers of some Ophrys species to their pollinators. Some male scoliid wasps such as *Campsoscolia ciliata* are more attracted to the Ophrys flowers' odours than to the odours of the female wasps, although they both attract the males with the same compounds. This is most likely a result of a higher amount of scent coming from the orchid flowers; female wasps tend to produce less scent to avoid attracting predators. Regardless of whether orchids use appearances, fragrances or both, they mimic the female pollinator for their own benefit.

References

- Edward G. Reekie; Fakhri A. Bazzaz (28 October 2005). Reproductive allocation in plants. Academic Press. pp. 99–. ISBN 978-0-12-088386-8. Retrieved 01, June 2020
- Knox, David; Schacht, Caroline. Choices in Relationships: An Introduction to Marriage and the Family. 11 ed. Cengage Learning; 2011-10-10 [cited 17 June 2013]. ISBN 9781111833220. p. 64–66
- Neuhof, Moran; Levin, Michael; Rechavi, Oded (26 August 2016). "Vertically and horizontally-transmitted memories – the fading boundaries between regeneration and inheritance in planaria". Biology Open: bio.020149. doi:10.1242/bio.020149. PMID 27565761. Retrieved 15, January 2020
- Helen Nilsson Sköld; Matthias Obst; Mattias Sköld & Bertil Åkesson (2009). "Stem Cells in Asexual Reproduction of Marine Invertebrates". In Baruch Rinkevich & Valeria Matranga. Stem Cells in Marine Organisms. Springer. p. 125. ISBN 978-90-481-2766-5
- Avise, J. (2008) Clonality: The Genetics, Ecology and Evolution of Sexual Abstinence in Vertebrate Animals. Oxford University Press. ISBN 019536967X ISBN 978-0195369670
- Christopher Ryan & Cacilda Jethá (2010). Sex at Dawn: The Prehistoric Origins of Modern Sexuality. Harper. ISBN 978-0-06-170780-3
- Lynn Margulis; Heather I. McKhann; Lorraine Olendzenski (2001). Illustrated glossary of protoctista: vocabulary of the algae, apicomplexa, ciliates, foraminifera, microspora, water molds, slime molds, and the other protoctists. Jones & Bartlett learn. ISBN 978-0-86720-081-2

- A. Jonathan Shaw (2000). "Population ecology, population genetics, and microevolution". In A. Jonathan Shaw & Bernard Goffinet. Bryophyte Biology. Cambridge: Cambridge University Press. pp. 379–380. ISBN 978-0-521-66097-6
- Rebecca Jordan-Young (September 2010). Brain Storm: The Flaws in the Science of Sex Differences. Harvard University Press. ISBN 978-0-674-05730-2
- Cordelia Fine (August 2010). Delusions of Gender: How Our Minds, Society, and Neurosexism Create Difference (1st ed.). W. W. Norton & Company. ISBN 978-0-393-06838-2

5

Understanding Ethology

The branch of biology that deals with the scientific and objective study of animal behavior is called ethology. It focuses on animal behavior under normal and hostile conditions and its relation to evolutionarily adaptive traits. Bunting, deimatic behavior, collective behavior, animal consciousness, animal communication, etc. all fall under its domain. This chapter has been carefully written to provide a better understanding of ethology and its related aspects.

Ethology is a branch of zoology concerned with the study of animal behavior. Ethologists take a comparative approach, studying behaviors ranging from kinship, cooperation, and parental investment, to conflict, sexual selection, and aggression across a variety of species. Today *ethology* as a disciplinary label has largely been replaced by behavioral ecology and evolutionary psychology. These rapidly growing fields tend to place greater emphasis on social relationships rather than on the individual animal; however, they retain ethology's tradition of fieldwork and its grounding in evolutionary theory.

The egg-rolling behavior of the greylag goose is a widely cited example of a *fixed-action pattern*, one of the key concepts used by ethologists to explain animal behavior

The study of animal behavior touches upon the fact that people receive joy from nature and also typically see themselves in a special role as stewards of creation. Behavior is one aspect of the vast diversity of nature that enhances human enjoyment. People are fascinated with the many behaviors of animals, whether the communication "dance" of honeybees, or the hunting behavior of the big cats, or the altruistic behavior of a dolphin. In addition, humans generally see themselves with the responsibility to love and care for nature.

The study of animal behavior also helps people to understand more about themselves. From an evolutionary point of view, organisms of diverse lineages are related through the process of descent with modification. From a religious point of view, human also stand as "microcosms of nature" (Burns 2006). Thus, the understanding of animals helps to better understand ourselves.

Ethologists engage in hypothesis-driven experimental investigation, often in the field. This combination of lab work with field study reflects an important conceptual underpinning of the discipline: behavior is assumed to be *adaptive*; i.e., something that makes it better suited in its environment and consequently improves its chances of survival and reproductive success.

Ethology emerged as a discrete discipline in the 1920s, through the efforts of Konrad Lorenz, Karl von Frisch, and Niko Tinbergen, who were jointly awarded the 1973 Nobel Prize in Physiology or Medicine for their contributions to the study of behavior. They were in turn influenced by the foundational work of, among others, ornithologists Oskar Heinroth and Julian Huxley and the American myrmecologist (study of ants) William Morton Wheeler, who popularized the term *ethology* in a seminal 1902 paper.

Important Concepts

One of the key ideas of classical ethology is the concept of fixed action patterns (FAPs). FAPs are stereotyped behaviors that occur in a predictable, inflexible sequence in response to an identifiable stimulus from the environment.

For example, at the sight of a displaced egg near the nest, the greylag goose *(Anser anser)* will roll the egg back to the others with its beak. If the egg is removed, the animal continues to engage in egg-rolling behavior, pulling its head back as if an imaginary egg is still being maneuvered by the underside of its beak. It will also attempt to move other egg-shaped objects, such as a golf ball, doorknob, or even an egg too large to have been laid by the goose itself.

Another important concept is *filial imprinting,* a form of learning that occurs in young animals, usually during a critical, formative period of their lives. During imprinting, a young animal learns to direct some of its social responses to a parent or sibling.

Despite its valuable contributions to the study of animal behavior, classical ethology also spawned problematic general theories that viewed even complex behaviors as genetically hardwired (i.e., *innate* or *instinctive*). Models of behavior have since been revised to account for more flexible decision-making processes.

Methodology

Tinbergen's Four Questions for Ethologists

The practice of ethological investigation is rooted in hypothesis-driven experimentation. Lorenz's collaborator, Niko Tinbergen, argued that ethologists should consider the following categories when attempting to formulate a hypothesis that explains any instance of behavior:

- Function: How does the behavior impact the animal's chance of survival and reproduction?
- Mechanism: What are the stimuli that elicit the response? How has the response been modified by recent learning?
- Development: How does the behavior change with age? What early experiences are necessary for the behavior to be demonstrated?
- Evolutionary history: How does the behavior compare with similar behavior in related

species? How might the behavior have arisen through the evolutionary development of the species, genus, or group?

The four questions are meant to be complementary, revealing various facets of the motives underlying a given behavior.

Using Fieldwork to Test Hypotheses

As an example of how an ethologist might approach a question about animal behavior, consider the study of hearing in an echolocating bat. A species of bat may use frequency chirps to probe the environment while in flight. A traditional neuroscientific study of the auditory system of the bat would involve anesthetizing it, performing a craniotomy to insert recording electrodes in its brain, and then recording neural responses to pure tone stimuli played from loudspeakers. In contrast, an ideal ethological study would attempt to replicate the natural conditions of the animal as closely as possible. It would involve recording from the animal's brain while it is awake, producing its natural calls while performing a behavior such as insect capture.

Key Principles and Concepts

Behaviors are Adaptive Responses to Natural Selection

Because ethology is understood as a branch of biology, ethologists have been particularly concerned with the evolution of behavior and the understanding of behavior in terms of the theory of natural selection. In one sense, the first modern ethologist was Charles Darwin, whose book *The Expression of the Emotions in Man and Animals* (1872) has influenced many ethologists. (Darwin's protégé George Romanes became one of the founders of comparative psychology, positing a similarity of cognitive processes and mechanisms between animals and humans.)

Note, however, that this concept is necessarily speculative. Behaviors are not found as fossils and cannot be traced through the geological strata. And concrete evidence for the theory of modification by natural selection is limited to microevolution—that is, evolution at or below the level of species. The evidence that natural selection directs changes on the macroevolutionary level necessarily involves extrapolation from these evidences on the microevolutionary level. Thus, although scientists frequently allude to a particular behavior having evolved by natural selection in response to a particular environment, this involves speculation as opposed to concrete evidence.

Animals use Fixed Action Patterns in Communication

As mentioned above, a *fixed action pattern (FAP)* is an instinctive behavioral sequence produced by a neural network known as the *innate releasing mechanism* in response to an external sensory stimulus called the *sign stimulus* or *releaser*. Once identified by ethologists, FAPs can be compared across species, allowing them to contrast similarities and differences in behavior with similarities and differences in form (morphology).

An example of how FAPs work in animal communication is the classic investigation by Austrian ethologist Karl von Frisch of the so-called "dance language" underlying bee communication. The dance is a mechanism for successful foragers to recruit members of the colony to new sources of nectar or pollen.

The honeybee's figure-eight dance is a fixed-action pattern that communicates information to other members of the group: the angle from the sun indicates the direction of a food source; the duration signifies its distance

Imprinting is a Type of Learning Behavior

Imprinting describes any kind of phase-sensitive learning (i.e., learning that occurs at a particular age or life stage) during which an animal learns the characteristics of some stimulus, which is therefore said to be "imprinted" onto the subject.

The best known form of imprinting is *filial imprinting,* in which a young animal learns the characteristics of its parent. Lorenz observed that the young of waterfowl such as geese spontaneously followed their mothers from almost the first day after they were hatched. Lorenz demonstrated how incubator-hatched geese would imprint on the first suitable moving stimulus they saw within what he called a critical period of about 36 hours shortly after hatching. Most famously, the goslings would imprint on Lorenz himself (more specifically, on his wading boots).

Sexual imprinting, which occurs at a later stage of development, is the process by which a young animal learns the characteristics of a desirable mate. For example, male zebra finches appear to prefer mates with the appearance of the female bird that rears them, rather than mates of their own type. *Reverse* sexual imprinting has also observed: when two individuals live in close domestic proximity during their early years, both are desensitized to later sexual attraction. This phenomenon, known as the *Westermarck effect,* has probably evolved to suppress inbreeding.

Relation to Comparative Psychology

In order to summarize the defining features of ethology, it might be helpful to compare classical ethology to early work in comparative psychology, an alternative approach to the study of animal behavior that also emerged in the early 20th century. The rivalry between these two fields stemmed in part from disciplinary politics: ethology, which had developed in Europe, failed to gain a strong foothold in North America, where comparative psychology was dominant.

Broadly speaking, comparative psychology studies general processes, while ethology focuses on adaptive specialization. The two approaches are complementary rather than competitive, but they do lead to different perspectives and sometimes to conflicts of opinion about matters of substance:

- Comparative psychology construes its study as a branch of psychology rather than as an outgrowth of biology. Thus, where comparative psychology sees the study of animal behavior in the context of what is known about human psychology, ethology situates animal behavior in the context of what is known about animal anatomy, physiology, neurobiology, and phylogenetic history.

- Comparative psychologists are interested more in similarities than differences in behavior; they are seeking general laws of behavior, especially relating to development, which can then be applied to all animal species, including humans. Hence, early comparative psychologists concentrated on gaining extensive knowledge of the behavior of a few species, while ethologists were more interested in gaining knowledge of behavior in a wide range of species in order to be able to make principled comparisons across taxonomic groups.
- Comparative psychologists focused primarily on lab experiments involving a handful of species, mainly rats and pigeons, whereas ethologists concentrated on behavior in natural situations.

Since the 1970s, however, animal behavior has become an integrated discipline, with comparative psychologists and ethological animal behaviorists working on similar problems and publishing side by side in the same journals.

Recent Developments in the Field

In 1970, the English ethologist John H. Crook published an important paper in which he distinguished *comparative ethology* from *social ethology*. He argued that the ethological studies published to date had focused on the former approach—looking at animals as individuals—whereas in the future ethologists would need to concentrate on the social behavior of animal groups.

Since the appearance of E. O. Wilson's seminal book *Sociobiology: The New Synthesis* in 1975, ethology has indeed been much more concerned with the social aspects of behavior, such as phenotypic altruism and cooperation. Research has also been driven by a more sophisticated version of evolutionary theory associated with Wilson and Richard Dawkins.

Furthermore, a substantial rapprochement with comparative psychology has occurred, so the modern scientific study of behavior offers a more or less seamless spectrum of approaches—from animal cognition to comparative psychology, ethology, and behavioral ecology. *Evolutionary psychology*, an extension of behavioral ecology, looks at commonalities of cognitive processes in humans and other animals as we might expect natural selection to have shaped them. Another promising subfield is *neuroethology*, concerned with how the structure and functioning of the brain controls behavior and makes learning possible.

Animal Behavior

Animal behaviour, the concept, broadly considered, referring to everything animals do, including movement and other activities and underlying mental processes. Human fascination with animal behaviour probably extends back millions of years, perhaps even to times before the ancestors of the species became human in the modern sense. Initially, animals were probably observed for practical reasons because early human survival depended on knowledge of animal behaviour. Whether hunting wild game, keeping domesticated animals, or escaping an attacking predator, success required intimate knowledge of an animal's habits. Even today, information about animal behaviour is of considerable importance. For example, in Britain, studies on the social organization and the ranging patterns of badgers (Meles meles) have helped reduce the spread of tuberculosis among cattle, and studies of sociality in foxes (Vulpes vulpes) assist in the development of models that predict how quickly rabies would spread should it ever cross the English Channel. Likewise in Sweden, where collisions involving moose (Alces alces) are among the most common traffic accidents in rural areas, research on moose behaviour has yielded ways of keeping them off roads and verges. In addition, investigations of the foraging of insect pollinators, such as honeybees, have led to impressive increases in agricultural crop yields throughout the world.

Konrad Lorenz being followed by greylag geese (Anser anser)

Even if there were no practical benefits to be gained from learning about animal behaviour, the subject would still merit exploration. Humans (Homo sapiens) are animals themselves, and most humans are deeply interested in the lives and minds of their fellow humans, their pets, and other creatures. British ethologist Jane Goodall and American field biologist George Schaller, as well as British broadcaster David Attenborough and Australian wildlife conservationist Steve Irwin, have brought the wonders of animal behaviour to the attention and appreciation of the general public. Books, television programs, and movies on the subject of animal behaviour abound.

The study of animal behavior is the scientific study of everything that animals do, whether the animals are single-celled organisms, invertebrates, fish, amphibians, reptiles, birds, humans, or other mammals. It involves the investigation of the relationship of animals to their physical environment as well as to other organisms, and includes topics such as how animals find and defend resources, avoid predators, choose mates and reproduce, and care for their young.

People who study animal behavior are typically trying to answer one or more of the following four kinds of questions about behavior (often referred to as "Tinbergen's four questions," after the animal behaviorist Niko Tinbergen who first described them):

- What is the cause of the behavior?
- What mechanism is underlying it?
- What "triggered" it just now?

- How did the behavior develop within the individual's lifetime?
- What function or functions does the behavior serve?
- How did the behavior evolve over time?

Answers to questions about the causes of behavior include both the external stimuli that affect behavior, and the internal hormonal and neural mechanisms that control behavior. Questions about the development of behavior focus on the ways in which behavior changes over the lifetime of the animal, and how these changes are affected by both genes and experience. Questions about the functions of behavior focus on both the behavior's immediate effect on the animal and on the behavior's adaptive value in helping the animal to survive or reproduce successfully in a particular environment. Finally, questions about the evolution of behavior focus on the origins of behavior patterns and how these change over generations.

Animal Sexual Behavior

The study of animal sexuality (and primate sexuality especially) is a rapidly developing field. It used to be believed that only humans and a handful of species performed sexual acts other than for procreation, and that animals' sexuality was instinctive and a simple response to the "right" stimulation (sight, scent). Current understanding is that many species that were formerly believed monogamous have now been proven to be promiscuous or opportunistic in nature; a wide range of species appear both to masturbate and to use objects as tools to help them do so; in many species animals try to give and get sexual stimulation with others where procreation is not the aim; and homosexual behaviour has now been observed among 1,500 species and in 500 of those it is well documented.

Mating Systems

In sociobiology and behavioural ecology, the term mating system is used to describe the ways in which animal societies are structured in relation to sexual behaviour. The mating system specifies which males mate with which females, and under what circumstances.

The following are some of the mating systems generally recognised in humans and other animals:

Monogamy: One male and one female have an exclusive mating relationship.

Polygamy: A single individual concurrently carries a relationship/mates with one or more of the opposite sex. Three types are recognized:

Polygyny: (the most common polygamous mating system in vertebrates so far studied): One male has an exclusive relationship with two or more females.

Polyandry: One female has an exclusive relationship with two or more males.

Polygynandry: Two or more individuals have an exclusive relationship with two or more individuals from the opposite sex; the numbers of males and females need not be equal, and in vertebrate species studied so far, there are usually fewer males.

Promiscuity: Any male and female will mate within the social group.

Monogamy: Monogamous Pairing in Animals

Zoologists and biologists now have solid evidence that monogamous pairs of animals are not always sexually exclusive. Many animals that form pairs to mate and raise offspring regularly engage in sexual activities with extra-pair partners. This includes previous exemplars such as swans. Sometimes these extra-pair sexual activities lead to offspring. Genetic tests frequently show that some of the offspring raised by a monogamous pair come from the female mating with an extra-pair male partner. These discoveries have led biologists to adopt new ways of talking about monogamy:

Social Monogamy

Social monogamy refers to a male and female's social living arrangement (e.g., shared use of a territory, behaviour indicative of a social pair, and/or proximity between a male and female) without inferring any sexual interactions or reproductive patterns. In humans, social monogamy equals monogamous marriage. Sexual monogamy is defined as an exclusive sexual relationship between a female and a male based on observations of sexual interactions. Finally, the term genetic monogamy is used when DNA analyses can confirm that a female-male pair reproduce exclusively with each other. A combination of terms indicates examples where levels of relationships coincide, e.g., sociosexual and sociogenetic monogamy describe corresponding social and sexual, and social and genetic monogamous relationships, respectively.

Whatever makes a pair of animals socially monogamous does not necessarily make them sexually or genetically monogamous. Social monogamy, sexual monogamy, and genetic monogamy can occur in different combinations.

Social monogamy is relatively rare in the animal kingdom. The actual incidence of social monogamy varies greatly across different branches of the evolutionary tree. Over 90 percent of avian species are socially monogamous.

This stands in contrast to mammals. Only 3 percent of mammalian species are socially monogamous, although up to 15 percent of primate species are socially monogamous. Social monogamy has also been observed in reptiles, fish, and insects.

Sexual monogamy is also rare among animals. Many socially monogamous species engage in extra-pair copulations, making them sexually non-monogamous. For example, while over 90% of birds are socially monogamous, "on average, 30 percent or more of the baby birds in any nest are sired by someone other than the resident male." Patricia Adair Gowaty has estimated that, out of 180 different species of socially monogamous songbirds, only 10 percent are sexually monogamous.

The incidence of genetic monogamy, determined by DNA fingerprinting, varies widely across species. For a few rare species, the incidence of genetic monogamy is 100 percent, with all offspring genetically related to the socially monogamous pair. But genetic monogamy is strikingly low in other species.

Polygamy

Polygamy is defined as a mating structure in which a single individual of one gender has exclusive access to several individuals of the opposite gender. It takes two main forms - polygyny and polyandry. As polygyny is the most common form of polygamy among vertebrates (including humans, to some extent), it has been studied far more extensively than polyandry.

In some species, notably those with harem-like structures, only one of a few males in a group of females will mate. Technically, polygyny in sociobiology and zoology is defined as a system in which a male has a relationships with more than one female, but the females are predominantly bonded to a single male.Should the active male be driven out, killed, or otherwise removed from the group, in a number of species the new male will ensure that breeding resources are not wasted on another male's young. The new male may achieve this in many different ways, including:

Competitive Infanticide

In lions and some monkeys, the new male will kill the offspring of the previous alpha male to cause their mothers to become receptive to his sexual advances since they are no longer nursing.

Harassment to Miscarriage

Amongst wild horses and baboons, the male will "systematically harass" pregnant females until they miscarry. Pheromone based spontaneous abortion. In some rodents such as mice, a new male with a different scent will cause females who are pregnant to spontaneously fail to implant recently fertilized eggs. This does not require contact; it is mediated by scent alone. It is known as the Bruce-Parkes effect.

Promiscuity

Two examples of systems in primates are promiscuous mating chimpanzees and bonobos. These species live in social groups consisting of several males and several females. Each female copulates with many males, and vice versa. In bonobos, the amount of promiscuity is particularly striking because bonobos use sex to alleviate social conflict as well as to reproduce.

Seasonal Nature of Animal Sexuality

Many animal species have specific mating (or breeding) seasons (seasonal breeding). These are often associated with changes to herd or group structure, and behavioural changes, including territorialism amongst individuals. These may be annual (e.g. wolves), biannual (e.g. dogs) or more frequently (e.g. horses). During these periods, females of most species are more mentally and physically receptive to sexual advances, a period scientifically described as estrous but commonly described as being "in season" or "in heat", but outside them animals still engage in sexual behaviours, and such acts as do occur are not necessarily harmful.

Interpretation of Animal Sexuality

The field of study of sexuality in non-human species has been a long standing taboo, with researchers either failing to observe or mis-categorizing and mis-describing sexual behaviour which does not meet their preconceptions.

More current research provides views such as that of the Natural History Museum at the University of Oslo, which in 2006 held an exhibition on animal sexuality:

Many researchers have described homosexuality as something altogether different from sex. They must realise that animals can have sex with who they will, when they will and without consideration to a researcher's ethical principles.

An example of overlooking behaviour relates to descriptions of giraffe mating:

When nine out of ten pairings occur between males, "every male that sniffed a female was reported as sex, while anal intercourse with orgasm between males was only 'revolving around' dominance, competition or greetings.

Sex for Pleasure

It is a common myth that animals do not (as a rule) have sex for pleasure, or alternatively that humans, pigs (and perhaps cats, dolphins and one or two species of primate) are the only species which do. This is sometimes formulated "animals mate only for reproduction".

Science cannot say at present conclusively what animals do or do not find "pleasurable", a question considered in more depth under Emotion in animals. The urban myth website Snopes.com considers this particular view in depth. Its conclusions are broadly that the statement is true, but only using a very specific definition of "sex for pleasure", in which sexual acts tied to a reproductive cycle or for which an alternative explanation can be asserted, are ignored, as is all sexual activity that does not involve penetration. Animals put themselves at risk to engage in sex, and as a result, most species have evolved sexual signals (usually scent and behaviour) to indicate the presence of receptive periods. During these, sex is sought, and outside these it is usually not sought (or is sought but not permitted). Snopes comments that this is not in fact a reflection of whether sex is pleasurable or not, but rather a reflection of whether individuals have sex at arbitrary times.

Types of Sexual Activity

Autoeroticism (Masturbation)

It appears that many animals, both male and female, masturbate, both when partners are available and otherwise.

For example, Petplace.com comments in its guide on assessing potential breeding stock purchases: "Masturbation is a normal behaviour in all stallions that does not reduce semen production or performance in the breeding shed"

Likewise a review from the University of Pennsylvania School of Veterinary Medicine says:

The behaviour known within the horse breeding industry as masturbation. This involves normal periodic erections and penile movements. This behavior, both from the descriptive field studies cited above and in extensive study of domestic horses, is now understood as normal, frequent behavior of male equids. Attempting to inhibit or punish masturbation, for example by tying a brush to the area of the flank underside where the penis rubs into contact with the underside, which is still a common practice of horse managers regionally around the world, often leads to increased masturbation and disturbances of normal breeding behaviour.

Castration does not prevent masturbation, as it is observed in geldings. Masturbation is common in both mares and stallions, before and after puberty.

The female porcupine has been observed to use a stick as a dildo, holding one end of a stick and walking around, straddling it as it bumped against the ground and vibrated against her genitalia.

Autoeroticism also occurs widely among animals, both male and female. A variety of creative techniques are used, including genital stimulation using the hand or front paw (primates, Lions), foot (Vampire Bats, primates), flipper (Walruses), or tail (Savanna Baboons), sometimes accompanied by stimulation of the nipples (Rhesus Macaques, Bonobos); auto-fellating or licking, sucking and/or nuzzling by a male of his own penis (Common Chimpanzees, Savanna Bonobos, Vervet Monkeys, Squirrel Monkeys, Thinhorn Sheep, Bharal, Aovdad, Dwarf Cavies); stimulation of the penis by flipping or rubbing it against the belly or in its own sheath (White-tailed and Mule Deer, Zebras and Takhi); spontaneous ejaculations (Mountain Sheep, Warthogs, Spotted Hyenas); and stimulation of the genitals using inanimate objects (found in several primates and cetaceans).

Many birds masturbate by mounting and copulating with tufts of grass, leaves or mounds of earth, and some mammals such as primates and Dolphins also rub their genitals against the ground or other surfaces to stimulate themselves.

Autoeroticism in female mammals, as well as heterosexual and homosexual intercourse (especially in primates), often involves direct or indirect stimulation of the clitoris. This organ is present in the females of all mammalian species and several other animal groups. and that:

Apes and Monkeys use a variety of objects to masturbate with and even deliberately create implements for sexual stimulation often in highly creative ways.

Petter Bøckman of the Natural History Museum at the University of Oslo commented (in respect of a 2006 exhibition on homosexuality in the animal kingdom) that: *Masturbation is common in the animal kingdom ... We have a Darwinist mentality that all animals only have sex to procreate. But there are plenty of animals who will masturbate when they have nothing better to do. Masturbation has been observed among primates, deer, killer whales and penguins, and we're talking about both males and females. They rub themselves against stones and roots. Orangutans are especially inventive. They make dildos of wood and bark.*

Oral sex

Animals of several species are documented as engaging in both autofellatio and oral sex. Although easily confused by lay-people, this is a separate and sexually oriented behaviour, distinct from non-sexual grooming or the investigation of scents.

Auto-fellatio or oral sex in animals is documented in goats, primates, hyaenas, bats and sheep.

Homosexual Behaviour - Homosexual Behavior in Animals

Mallard Ducks *Anas platyrhynchos* form male-female pairs only until the female lays eggs, at which time the male leaves the female. Mallards have rates of male-male sexual activity that are unusually high for birds, in some cases, as high as 19% of all pairs in a population. The presence of same-sex sexual behaviour was not scientifically observed on a large scale until recent times. Homosexual behaviour does occur in the animal kingdom outside humans, especially in social species, particularly in marine birds and mammals, monkeys, and the great apes. Homosexual behaviour has been observed among 1,500 species, and in 500 of those it is well documented.

To turn the approach on its head: No species has been found in which homosexual behaviour has not been shown to exist, with the exception of species that never have sex at all, such as sea urchins and aphis. Moreover, a part of the animal kingdom is hermaphroditic, truly bisexual. For them, homosexuality is not an issue.

Georgetown University professor Janet Mann has specifically theorised that homosexual behaviour, at least in dolphins, is an evolutionary advantage that minimizes intraspecies aggression, especially among males.

Male penguin couples have been documented to mate for life, build nests together, and to use a stone as a surrogate egg in nesting and brooding. In 2004, the Central Park Zoo in the United States replaced one male couple's stone with a fertilized egg, which the couple then raised as their own offspring. German and Japanese zoos have also reported homosexual behaviour among their penguins. This phenomenon has also been reported at Kelly Tarlton's Aquarium in Auckland, New Zealand. "Humans have created the myth that sexuality can be justified only by reproduction, which by definition limits it to hetero sex," says Michael Bronski, author of The Pleasure Principle: Culture, Backlash, and the Struggle for Gay Freedom. "But here is an animal society that uses homosexuality to improve its social life."

Mounting of one female by another is common among cattle.

After studying the Bonobo primates for his book Bonobo: The Forgotten Ape, primatologist Frans de Waal, a professor of psychology at Emory University in Atlanta, says that such expressions of intimacy are consistent with the homosexual behaviour of what he terms "the erotic champions of the world." "Same-sex, opposite-sex — bonobos just love sex play," de Waal said in an interview. "They have so much sex, it gets boring.

"Homosexual behaviour in male sheep (found in 6-10% of rams) is associated with variations in cerebral mass distribution and chemical activity. A study reported in Endocrinology concluded that biological and physiological factors are in effect. These findings are similar to human findings reported by Simon LeVay.

Approximately eight percent of [male] rams exhibit sexual preferences [that is, even when given a choice] for male partners (male-oriented rams) in contrast to most rams, which prefer female partners (female-oriented rams). We identified a cell group within the medial preoptic area/anterior hypothalamus of age-matched adult sheep that was significantly larger in adult rams than in ewes.

Male bighorn sheep are divisible into two kinds: the typical males among whom homosexual behaviour, including intercourse, is common and "effeminate sheep", or "behavioural transvestites", which are not known to engage in homosexual behaviour.

Cross Species Sex

While it is commonly believed that animal sexuality is instinctive and thus somewhat mechanistic, research regularly records that many animals are sexual opportunists, partaking in sexual relations with individuals of visibly distinct species. This is more visible in domesticated species and animals in captivity, as domestication commonly selects for increased breeding rate (and so an accelerated breeding cycle has commonly arisen in domesticated species over the centuries), and also because these species are more easily observed by humans. Nevertheless, animals have been observed in the wild to attempt sexual activity with other species or indeed inanimate objects. Attempts by wild moose to obtain sex from domestic horses are apparently well known by wildlife specialists.

In the wild, where observation is harder, genetic studies have shown a "large number" of inter-species hybrids, and other investigations describe productive and non-productive inter-species mating as a "natural occurrence".Recent genetic evidence strongly suggesting this has occurred even within the history of the human species, and that early humans often had sexual activity with other primate species, is considered below.

Hybrid offspring can result from two organisms of distinct but closely related parent species, although the resulting offspring is not always fertile. According to the definition of a species, if two organisms cannot or will not mate and produce a fertile offspring, they are different species. The mule, for example (a horse/donkey cross) is normally sterile, whilst the liger (lion/tiger cross) has fertile females and sterile males. Novosibirsk zoo director Rostislav Shilo says on the liger (born in its zoo); "It's just that the lion and the tiger live in neighboring caves in the Novosibirsk zoo, and got used to each other. It's practically impossible in the wild".

Due to the difficulties of observation, interspecies sex of this kind between two top-level predators, occurring in the wild, was only conclusively documented with the finding of a grizzly-polar bear hybrid in April 2006. Again, as with lions and tigers, the two species would normally not share enough common territory to provide adequate opportunity for much cross-species sexual activity.

Animal sexual advances on, and attempted interactions with, humans and other species, have been documented by ethologists such as Kohler, Gerald Durrell and Desmond Morris, as well as authoritative researchers such as Birute Galdikas who studied orangutans in Borneo. Philosopher and animal welfare activist Peter Singer reports: *While walking through the camp with Galdikas, my informant was suddenly seized by a large male orangutan, his intentions made obvious by his erect penis. Fighting off so powerful an animal was not an option, but Galdikas called to her companion not to be concerned, because the orangutan would not harm her, and adding, as further reassurance, that 'they have a very small penis' ... though the orangutan lost interest before penetration took place.*

Prostitution

In some penguin species, the females, even when in a committed relationship, will exchange sexual favours with strange males for the pebbles they need to build their nests. Prostitution was also observed among chimpanzees, who trade food for sex.

Sexual Fetishes

Although not often reported, animals, or primates at the least, are able to sexualize inanimate objects similar to the way human beings sexualize the objects of their sexual fetishes. Not only will an animal that has a habitual object for masturbation sometimes appear to sexualize that object, primates have generalized further to sexualize kinds of objects for which no instinctual or prior sexual connection exists.

Thus Gabriel, a chimpanzee at the Southwest National Primate Research Center, is said to have a shoe fetish (or possibly a leather fetish) according to caretaker Bert Barrera, and it is reported that a male chimpanzee raised in captivity developed a bit of a shoe fetish, masturbating obsessively by rubbing his caretaker's leather boot.

The sexualization of objects or locations is also well recognized in the breeding world. So for example, stallions may often 'drop' (become sexually aroused) upon visiting a location where they have been allowed to have sex before, or upon seeing a stimulus previously associated with sexual activity such as an artificial vagina.

In this case however, the primary structure is Pavlovian conditioning, and the fetishistic association is due to a conditioned response (or association) formed with a distinctive 'reward'. Human fetishism can also be traced back to similar or near-identical conditioning: likewise based upon the Pavlovian association between an erotic sensation or anticipation, and objects which become mentally associated with that activity.

Sexual Imagery Viewing

A study by Platt, Khera and Deaner at Duke University North Carolina (reported in Current Biology and online here), showed that male monkeys will give up privileges (in this case, juice, which is highly valued), to be allowed to see a female monkey's hindquarters.

Deaner and his team reported that monkeys would take a juice cut to look at powerful males' faces or the perineum of a female, but to persuade the monkeys to stare at subordinate males, the researchers had to bribe them with larger drinks. "Virtually all [male] monkeys will give up juice to see female hindquarters ... they really value the images."

The researchers stress that in monkey society, such behaviours have great social utility and we should therefore not simply reach the conclusion that "monkeys enjoy pornographic pictures". There is no evidence at this point that viewable pictures or movies of sexual activity are valued for their sexual enjoyment, although as noted above (Masturbation), there are reports that watching sex in real life may have such an effect. The subject of animals and sexual imagery is not yet well researched.

Problems with encouraging pandas to mate in captivity have been very common. However, showing young male pandas "panda pornography" is widely credited with a recent population boom among pandas in zoos.

Coercive Sex

Controversial interpretations and implications aside, sex in a forceful or apparently coercive context has also been documented in a variety of species. A notable example is bottlenose dolphins, where at times, a pod of bachelor males will 'corner' a female. Furthermore, in a zoo where it is common practice to put newly captured dolphins in with dolphins who are established in their enclosures, other species of dolphin are never put in together with bottlenoses because the bottlenose dolphins frequently torment and rape them. The behaviour is also common in some arachnids (spiders), notably those whose females eat the males during sex if not tricked with food and/or tied down with threads, and in some herbivorous herd species or species where males and females are very different in size, where the male dominates sexually by sheer force and size.

Some species of birds appear to combine sexual intercourse with apparent violent assault; these include ducks,geese, and white-fronted bee-eaters. According to Emlen and Wrege (1986) forced copulations occur in this socially nesting species, and females must avoid the unwelcome attention of males as they emerge from their nest burrows or they are forced to the ground and mated with. Apparently, such attacks are made preferentially on females who are laying and who may thus mother their offspring as a result.

In 2007, research suggested that in the Acilius genus of water beetles (also known as "diving beetles"), an "evolutionary arms race" between the genders means that there is no courtship system for these beetles. "It's a system of rape. But the females don't take things quietly. They evolve counter-weapons." Cited mating behaviours include males suffocating females underwater till exhausted, and allowing only occasional access to the surface to breathe for up to six hours (to prevent them breeding with other males), and females which have a variety of body shapings (to prevent males from gaining a grip). Foreplay is "limited to the female desperately trying to dislodge the male by swimming frantically around."

Charles Siebert reports in his New York Times article Elephant Crackup? that: Since the early 1990's, for example, young male elephants in Pilanesberg National Park and the Hluhluwe-Umfolozi Game Reserve in South Africa have been raping and killing rhinoceroses; this abnormal behaviour, according to a 2001 study in the journal Pachyderm, has been reported in "a number of reserves" in the region.

Sex between Adults and Juveniles

It has also been recorded that certain species of mole will impregnate newborns of their own species. It is not clear if this is forceful or not. Similarly, the male stoat (Mustela erminea) will mate with infant females of their species. This apparently is a natural part of their reproductive biology - there is a delayed gestation period, so these females give birth the following year when they are fully grown.

A male spotted hyena which attempted to mate with a female which succeeded in driving it off, eventually turned to its ten-month-old cub, repeatedly mounting it and ejaculating on it. The cub sometimes ignored this and sometimes struggled 'slightly as if in play'. The mother did not intervene.

Infants and children in Bonobo societies are often involved in sexual behaviour.

Sexual Cannibalism

Sexual cannibalism, which has been documented in arachnids, insects and amphipods, is a phenomenon in which a female organism kills and consumes the male before, during, or after copulation. Although it does confer some known advantages to reproduction, whether or not the male is complicit has not been scientifically determined.

Necrophilia

Necrophilia in animals is where a living animal engages in a sexual act with a dead animal. In one of the most well-known examples, Kees Moeliker of the Rotterdam Natural History Museum, Netherlands observed sexual activities outside his office between a live duck and a dead one. Two male mallards which Moeliker believed were engaged in rape flight, a common motif in duck sexual behaviour, collided with his window. "*When one died the other one just went for it and didn't get any negative feedback—well, didn't get any feedback,*" according to Moeliker, who described the event as "homosexual necrophilia." The case was reported scientifically in Deinsea 8-2001, along with photos, and earned Moeliker an Ig Nobel Prize in biology, awarded for research that cannot or should not be reproduced.

Additionally, male cane toads have been documented (in Cane Toads: An Unnatural History) engaging in copulation with dead toads and inanimate objects.

Notes on specific species:

Bonobos

The Bonobo, which has a matriarchal society, is a fully bisexual species — both males and females engage in sexual behaviour with the same and the opposite sex, with females being particularly noted for engaging in sexual behaviour with each other and at up to 75% of sexual activity being nonreproductive. Primatologist Frans de Waal believes that Bonobos use sexual activity to resolve conflict between individuals. Sexual activity occurs between almost all ages and sexes of Bonobo societies.

Birds

Some black swans of Australia form sexually active male-male mated pairs and steal nests, or form temporary threesomes with females to obtain eggs, driving away the female after she lays the eggs. More of their cygnets survive to adulthood than those of different-sex pairs possibly due to their superior ability to defend large portions of land.

In early February 2004 the New York Times reported that a male pair of chinstrap penguins named Roy and Silo in the Central Park Zoo in New York City were partnered and had successfully hatched a female chick from an egg. Other penguins in New York have also been reported to be forming same-sex pairs.

Zoos in Japan and Germany have also documented male penguin couples. The couples have been shown to build nests together and use a stone to replace an egg in the nest. Researchers at Rikkyo University in Tokyo, found twenty such pairs at sixteen major aquariums and zoos in Japan. Bremerhaven Zoo in Germany attempted to break up the male couples by importing female penguins from Sweden and separating the male couples; they were unsuccessful. The zoo director stated the relationships were too strong between the couples.

Recently, a mated pair of swans in Boston were found to both be female. They too had attempted to raise eggs together. Studies have shown that ten to fifteen percent of female western gulls in some populations in the wild prefer other females.

As many as 19% of Mallard pairs in a given population have been observed to consist of male-male homosexuals.

Lizards

Whip-tailed lizard females have the ability to reproduce through parthenogenesis and as such males are rare and sexual breeding non-standard. Females engage in sexual behaviour to stimulate ovulation, with their behaviour following their hormonal cycles; during low levels of estrogen, these (female) lizards engage in "masculine" sexual roles. Those animals with currently high estrogen levels assume "feminine" sexual roles.

Lizards that perform the courtship ritual have greater fecundity than those kept in isolation due to an increase in hormones triggered by the sexual behaviours. So, even though asexual whiptail lizards populations lack males, sexual stimuli still increase reproductive success.

From an evolutionary standpoint these females are passing their full genetic code to all of their offspring rather than the 50% of genes that would be passed in sexual reproduction. Certain species of gecko also reproduce by parthenogenesis.

Flatworm

Penis fencing is a mating behaviour engaged in by certain species of flatworm, such as Pseudobiceros bedfordi. Species which engage in the practice are hermaphroditic, possessing both eggs and sperm-producing testes.

The species "fence" using two-headed dagger-like penises which are pointed, and white in color. One organism inseminates the other. The sperm is absorbed through pores in the skin, causing fertilization.

Sheep

An October 2003, study by Dr. Charles E. Roselli et al. (Oregon Health & Science University) states that homosexuality in male sheep (found in eight percent of rams) is associated with a region in the rams' brains which the authors call the "ovine Sexually Dimorphic Nucleus" (oSDN) which is half the size of the corresponding region in other male sheep.

However, some view this study to be flawed in that the determination of homosexuality within the sheep, (sample population of twenty-seven for the study), was to have animals who were unable to mount female ewes placed in a cage with two stanchioned males and two unstanchioned females (that is, the males could not move or struggle while the females could). Given the aggressive nature of the sheep copulation, the uneven treatment of males and females, many see this as simply evidence that the sheep in question were unable to be aggressive enough to mount females. Some say that the results were situational sexuality, unlike the bonds seen in human homosexuality.

The scientists found that, "The oSDN in rams that preferred females was significantly larger and contained more neurons than in male-oriented rams and ewes. In addition, the oSDN of the female-oriented rams expressed higher levels of aromatase, a substance that converts testosterone to estradiol, an estrogen hormone believed to facilitate typical male sexual behaviours. Aromatase expression was no different between male-oriented rams and ewes."

"The dense cluster of neurons that comprise the oSDN express cytochrome P450 aromatase. Aromatase mRNA levels in the oSDN were significantly greater in female-oriented rams than in ewes, whereas male-oriented rams exhibited intermediate levels of expression." These results suggest that "...naturally occurring variations in sexual partner preferences may be related to differences in brain anatomy and its capacity for estrogen synthesis." As noted previously, given the potential unagressiveness of the male population in question, the differing aromatase levels may also have been evidence of aggression levels, not sexuality. The results of this study have not been confirmed by others.

Spotted Hyena

The female Spotted Hyena has a unique urinary-genital system, closely resembling the penis of the male, called a pseudo-penis. The family structure is matriarchal and dominance relationships with strong sexual elements are routinely observed between related females.

They are notable for using visible sexual arousal as a sign of submission and not dominance, in males as well as females (females have a sizable erectile clitoris), to the extent that biologist Robert Sapolsky speculates that in order to facilitate this, their sympathetic and parasympathetic nervous systems may be partially reversed in respect to their reproductive organs.

Bottlenose Dolphins

Bottlenose Dolphin males have been observed working in pairs to follow and/or restrict the movement of a female for weeks at a time, waiting for her to become sexually receptive. The same pairs have also been observed engaging in intense sexual play with each other.

Janet Mann, a professor of biology and psychology at Georgetown University, argues that the common same-sex behaviour among male dolphin calves is about bond formation and benefits the species evolutionarily. They cite studies that have shown the dolphins later in life are bisexual and the male bonds forged from homosexuality work for protection as well as locating females with which to reproduce.

In 1991 an English man was prosecuted for allegedly having sexual contact with a dolphin. The man was found not guilty after it was revealed at trial that the dolphin was known to tow bathers through the water by hooking its large penis around them.

Seahorses

Seahorses, long upheld as monogamous and mating for life, are identified as "promiscuous, flighty, and more than a little bit gay" according to research published in 2007.

Scientists at 15 aquariums studied 90 seahorses of 3 species. Of 3168 sexual encounters, 37% were same sex acts. Flirting was common (up to 25 potential partners a day of both genders); only one species (the British Spiny Seahorse) included faithful representatives, and for these 5 of 17 were faithful, 12 were not. Bisexuality was widespread and considered "both a great surprise and a shock", with big bellied seahorses of both genders not showing partner preference. 1986 contacts were male-female, 836 were female-female and 346 were male-male.

Lions

Male lions often lead their social groups jointly with one or more of their brothers. To ensure loyalty, the male co-leaders will "strengthen the bonds" by often having sex with each other.

Horses

Anecdotal evidence suggest that some horses have environment or appearance preferences when selecting mates. There is also anecdotal evidence of limited bisexual behaviour in some stallions, although there is no conclusive scientific confirmation. The anecdotal evidence claims this is most likely to occur in a single isolated group, with no access to mares.

Humboldt Penguins

In 2009 at a zoo in Bremerhaven, Germany, two male adult humboldt penguins adopted an egg that had been abandoned by its biological parents. After the egg hatched, the two penguins raised, protected, cared for, and fed the chick in the same manner that heterosexual penguins raise their own biological offspring.

Role in Discussion of Human Sexuality

Information about animal sexuality frequently arises as a persuasive device in arguments regarding human sexuality. Originally, the lack of documented animal sexual behaviour deviant from heterosexual sexual monogamy was used to argue that the dominant heterosexual monogamy of most modern human societies is more natural and acceptable. Likewise, the lack of documented sex between animals for the purpose of pleasure was used to promote the moral standard of reserving sex primarily for procreation. Proponents of alternate sexuality attribute this early lack of documented evidence to an observer bias in researchers, who, they argue, tended to interpret sexual behaviour inconsistent with their values as other behaviour.

With increasing published evidence of different types of sexual behaviour between animals, arguments for heterosexual monogamy in human society have moved towards characterizing these behaviours as resulting from differences between humans and animals, and in particular on ambiguity in motivation and subjective experience in animals, which is difficult to study. Arguments identifying human and animal behaviour are characterized as anthropomorphism, and in some cases an opposite observer bias is attributed to researchers. Supporters of alternate sexuality embrace the new research as confirmation of the naturalness of alternate sexual behaviour and evidence of its long-term feasibility and utility.

Bunting

Bunting is a form of animal behavior, often found in cats, in which the animal butts or rubs their head against other things, including people. Bunting as a behaviour can be viewed as a variation of scent rubbing. This is when an animal, typically a carnivore, will rub its back on a scent such as prey or the urine of a conspicific. Evolutionarily speaking, scent rubbing is the oldest form of scent communication and bunting is a derivative of this behaviour. Rolling in the scent of another animal was an adaptation to camouflage the scent of a predator or outside male in order to get closer to mates. Bunting is generally considered to be a form of territorial scent-marking behaviour, where the cat rubs the scent glands on their cheeks and forehead on the object being marked. After a display of aggression, a cat will begin bunting objects in that immediate area as a form of territorial display toward a rival cat. Bunting and allorubbing are also part of feral cat behavior within colonies. An elaborate ritual which can take several minutes, two cats will rub along the side and tail of the other cat. This behaviour in domestic cats involves a system of hierarchy and may have evolved as a way to channel aggression where the costs of a conflict is too high. Cats also use bunting as a way to familiarize themselves to their environment and the pheromones secreted work to ease the cat's anxieties about an unfamiliar area.

A cat demonstrating bunting behaviour on a dog

Bunting is a normal animal behavior, and should be distinguished from head pressing, which is abnormal and typically a sign of illness.

Development

The practice of bunting stems from the behaviour that arises when kittens are very young and seek stimulation from their mother by rubbing and kneading. This behaviour develops throughout the animals life and is not only found in cats. It has also been found in other carnivorous mammals and some ungulates. Bunting in ungulates has no olfactory function but may have a role in the weaning of young. When a juvenile is nursing from its mother, it will bunt the udder with its head. This is to stimulate milk production or "let down" and causes some pain to the mother when the bunting movement is frequently performed. Overtime, there is an increase in the amount of times the udder is bunted by the young. This causes the mother to react in an attempt to prevent further pain. This reaction of the mother can be any form of defensive behaviour from a nipping bite, moving away from the young, or a jab of her horns. Bunting in domesticated and wild cat species has olfactory roots and has a range of uses which include, but are not limited to, mother-young association bonds, greeting/welcoming of kin, diffusing potential aggression in social environments, and distributing scent on areas to cultivate familiarity.

Bunting in other Species

Cattle

Bunting behaviour is a display of aggression in cattle. When two cattle are rivaling each other, they will often use bunting as a form of defense. Cattle will attempt to bunt the rival cattle with the goal of bunting their head under the hind legs of the animal. This occurs when one cattle shows submission during the final moments of a feud, and this specific behavior is calling clinching. The behaviour of bunting within cattle is first observable in calves. As a form of play fighting, a young calf will bunt the flank of its mother. A newborn calf will bunt the mother's udder and this stimulates milk flow. It has been found that when calves are taken from their mothers and raised artificially, the calf will attempt to bunt the artificial teat when milk is not being produced quick enough.

Horses

Horses will rub the bottom of their jaw onto others as a form of greeting. As seen in cattle, horse foals will bunt the mother's udder in order to stimulate milk production. Another example of bunting is when a dam experiences pain while a foal is suckling. The dam will proceed to bunt the foal with her head, non-aggressively, to prevent further discomfort. Many foals will play fight with one and other; displaying the bunting behaviour. The foal will push its head against another foal's body in an attempt to knock the other off-balance.

Lions

Lions display bunting as a form of greeting when met with other pride members

Similarly to domesticated cats, lions also use bunting as a form of greeting and territorial marking. Lions will often greet each other with this head bunting behavior when returning to a pride after a hunt. In the early stages of life, cubs procure stimulation from their mother as she cleans them by rubbing and licking them. This behaviour carries throughout their lives and bunting remains a primal source of interaction between adults as it stimulates a familiar interaction between kin.

Deimatic Behavior

Deimatic behaviour, threat display, or startle display in animals means any pattern of behaviour, such as suddenly displaying conspicuous eyespots, to scare off or momentarily distract a predator, thus giving the prey animal an opportunity to escape. The term deimatic or dymantic originates from the Greek δειματόω (deimatóo), meaning "to frighten".

Deimatic display: Callistoctopus macropus generates a bright brownish red colour with white oval spots when disturbed

Deimatic display occurs in widely separated groups of animals, including moths, butterflies, mantises and phasmids among the insects. In the cephalopods, different species of octopuses, squids, cuttlefish and the paper nautilus are deimatic.

Displays are classified as deimatic or aposematic by the responses of the animals that see them. Where predators are initially startled but learn to eat the displaying prey, the display is classed as deimatic, and the prey is bluffing; where they continue to avoid the prey after tasting it, the display is taken as aposematic, meaning the prey is genuinely distasteful. However, these categories are not entirely mutually exclusive. It is possible for a behaviour to be both deimatic and aposematic, if it both startles a predator and indicates the presence of anti-predator adaptations.

Vertebrates including several species of frog put on warning displays; some of these species have poison glands. Among the mammals, deimatic displays are found in species with strong defences, such as in foul-smelling skunks and spiny porcupines. Such displays are often combined with warning coloration. Thus these displays in both frogs and mammals are at least in part aposematic.

In Insects

Deimatic displays are made by insects including the praying mantises (Mantidae) and stick insects (Phasmatodea). While undisturbed, these insects are usually well camouflaged. When disturbed by a potential predator, they suddenly reveal their hind wings, which are brightly coloured. In mantises, the wing display is sometimes reinforced by showing brightly coloured front legs, and accompanied by a loud hissing sound created by stridulation. For example, the grasshopper

Phymateus displays red and yellow areas on its hind wings; it is also aposematic, producing a distasteful secretion from its thorax. Similarly the threat display of the walking stick phasmid (Peruphasma schultei) is not a bluff: the insect sprays defensive dolichodial-like monoterpene chemical compounds at attackers.

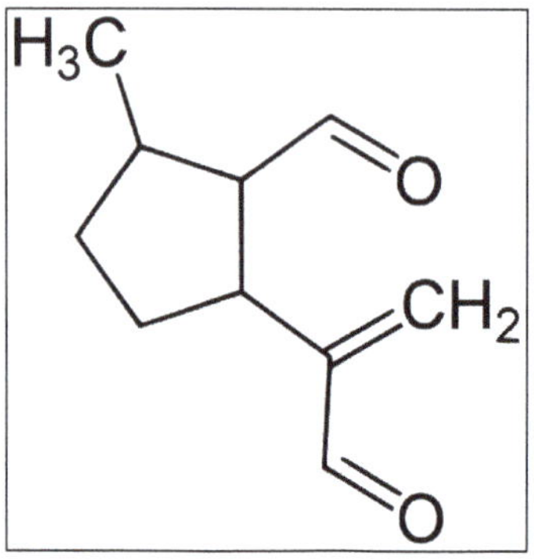

Threat displays are not always bluff. Some stick insects spray the monoterpene chemical dolichodial when attacked

A puss moth (Cerura vinula) caterpillar displaying its two flagella on its tail and red patches on its head. If the threat doesn't retreat, the caterpillar can fire formic acid from its flagella

Among moths with deimatic behaviour, the eyed hawkmoth (Smerinthus ocellatus) displays its large eyespots, moving them slowly as if it were a vertebrate predator such as an owl. Among butterflies, the peacock butterfly Aglais io is a cryptic leaf mimic with wings closed, but displays 4 conspicuous eyespots when disturbed, in a display effective against insectivorous birds (flycatchers).

An experiment by the Australian zoologist A.D. Blest demonstrated that the more an eyespot resembled a real vertebrate eye in both colour and pattern, the more effective it was in scaring off insectivorous birds. In another experiment using peacock butterflies, Blest showed that when the conspicuous eyespots had been rubbed off, insectivorous birds (yellow buntings) were much less effectively frightened off, and therefore both the sudden appearance of colour, and the actual eyespot pattern, contribute to the effectiveness of the deimatic display.

Some noctuid moths, such as the large red underwing (Catocala nupta), are cryptic at rest, but display a flash of startlingly bright colours when disturbed. Others, such as many species of genus Speiredonia and Spirama, look threatening while at rest. Also saturniid moths of the genera Attacus and Rothschildia display snake heads, but not from the frontal position.

Many arctiid moths make clicks when hunted by echolocating bats; they also often contain unpalatable chemicals. Some such as dogbane tiger moths (Cycnia tenera) have ears and conspicuous coloration, and start to make clicks when echolocating bats approach. An experiment by Canadian zoologists John M. Ratcliffe and James H. Fullard, using dogbane tiger moths and northern long-eared bats (Myotis septentrionalis), suggests the signals in fact both disrupt echolocation and warn of chemical defence. The behaviour of these insects is thus both deimatic and aposematic.

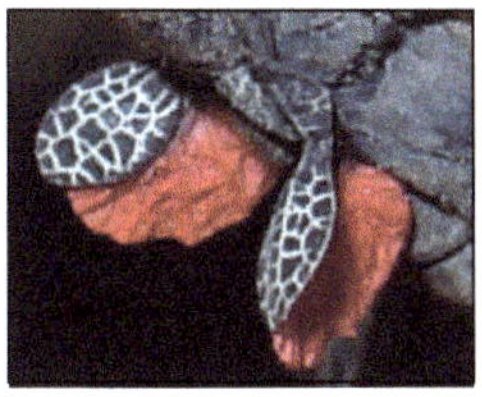

Deimatic display of the phasmid Peruphasma schultei

An adult female Mediterranean mantis, Iris oratoria, in threat pose

Speiredonia spectans resting mimicking a brooding head

Pseudocreobotra wahlbergii flashing its wings in deimatic pose

Peacock butterfly displays startling eyespots.

Threat pose of the phasmid Haaniella dehaanii

Female eyed hawk-moth, Smerinthus ocellatus, mounted to show the large eyespots

Peacock butterfly, Inachis io is a cryptic leaf mimic when its wings are closed

Spirama helicina displaying the face of a snake while at rest

A fine large "Phasma" illustrated by George Robert Gray, showing cryptic resting pose and dramatic wing flash

In Arachnids

Both spiders and scorpions are venomous, so their threat displays can be considered generally aposematic. However, some predators such as hedgehogs and spider-hunting wasps (Pompilidae) actively hunt arachnids, overcoming their defences, so when a hedgehog is startled by, for instance, the sounds made by a scorpion, there is reason to describe the display as deimatic.

Spiders make use of a variety of different threat displays. Some such as Argiope and Pholcus make themselves and their webs vibrate rapidly when they are disturbed; this blurs their outline and perhaps makes them look larger, as well as more difficult to locate precisely for an attack. Mygalomorphae spiders such as tarantulas exhibit deimatic behaviour; when threatened, the spider rears back with its front legs and pedipalps spread and fangs bared. Some species, such as the dangerous Indian ornamental tree spider (Poecilotheria regalis) have bright colouring on the front legs and mouthparts which are shown off in its threat display when it "rears up on its hind legs, and brandishes the fore limbs and palpi in the air".

Scorpions perform non-bluffing threat displays, as they have powerful defences, but various predators still eat them. When provoked, they spread their pincers and in some cases raise their abdomens, their tails standing near-erect with the sting ready for immediate use. Some scorpions in addition produce deimatic noises by stridulating with the pedipalps and first legs.

Aposematic threat display of Brazilian tarantula

Belly of the spider Poecilotheria regalis. The bright yellow forelegs are used in deimatic displays

Scorpion's threat display with pincers spread wide, abdomen raised to present sting

In Cephalopods

Deimatic behaviour is found in cephalopods including the common cuttlefish Sepia officinalis, squid such as the Caribbean reef squid (Sepioteuthis sepioidea) and bigfin reef squid (Sepioteuthis lessoniana), octopuses including the common octopus Octopus vulgaris and the Atlantic white-spotted octopus (Octopus macropus), and the paper nautilus (Argonauta argo). Deimatic cephalopod displays involve suddenly creating bold stripes, often reinforced by stretching out the animal's arms, fins or web to make it look as big and threatening as possible.

For example, in the common cuttlefish the display consists of flattening the body, making the skin pale, showing a pair of eyespots on the mantle, dark eye rings, and a dark line on the fins, and dilating the pupils of the eyes. The common octopus similarly displays pale skin and dark eye rings with dilated pupils, but also curls its arms and stretches out the web between the arms as far as possible, and squirts out jets of water. Other octopuses such as Atlantic white-spotted octopus turn bright brownish red with oval white spots all over in a high contrast display. The paper nautilus can rapidly change its appearance: it suddenly withdraws the shining iridescent web formed by its first pair of arms from its shell.

In Vertebrates

Frilled lizard faces predators, making itself look big with head frills, raising its body and waving its tail

Among vertebrates, the Australian frilled lizard (Chlamydosaurus kingii) has a startling display in which wide semicircular frills on either side of the head are fanned out; the mouth is opened wide exposing the gape; the tail is waved over the body, and the body is raised, so that the animal appears as large and threatening as possible.

Frogs such as Physalaemus nattereri, Physalaemus deimaticus, and Pleurodema brachyops have a warning display behaviour. These animals inflate themselves with air and raise their hind parts to appear as large as possible, and display brightly coloured markings and eyespots to intimidate predators. Seven species of frogs in the genus Pleurodema have lumbar glands (making the animals distasteful, so in their case the display is likely aposematic); these glands are usually boldly contrasted in black as a further warning.

Non-bluffing (aposematic) displays occur in mammals which possess powerful defences such as spines or stink glands, and which habitually warn off potential predators rather than attempting escape by running. The lowland streaked tenrec (Hemicentetes semispinosus) raises the spines on its head and back when confronted by a predator, and moves its head up and down. Porcupines such as Erethizon erect their long sharp quills and adopt a hunched, head-down posture when a predator is nearby. The spotted skunk (Spilogale putorius) balances on its front legs, its body raised vertically with its bold pelage pattern conspicuously displayed, and its tail (near the scent glands) raised and spread out.

Namaqua chameleon showing threat display with dewlap

Colombian four-eyed frog, Pleurodema brachyops

Euasian eagle owl, Bubo bubo, erects the feathers on its neck to make itself appear larger

Lowland streaked tenrec, Hemicentetes semispinosus erects spines on head and body when threatened

Striped skunk, Mephitis mephitis, displays prominent lighter markings against black, with raised bushy tail, honestly advertising its squirting scent glands

Domestic cat, Felis catus, hisses, arches its back, raises its fur to appear larger and bares its teeth in its genuine warning display

Deimatic or Aposematic?

In a study of the rattling made by rattlesnakes of different species, the Canadian zoologists Brock Fenton and Lawrence Licht found that the sounds are always similar: they have rapid onset (starting suddenly, and reaching full volume in a few milliseconds); they consist of a "broadband" mixture of frequencies between 2 kHz and 20 kHz, with little energy either in the ultrasonic (above 20 kHz) or in the rattlesnakes' hearing range (below 700 Hz); and the frequencies do not change much with time (the rattling after 2 minutes having a similar spectrum to that at onset). There was no clear difference in the sounds made by the different species measured: Crotalus horridus, Crotalus adamanteus, Crotalus atrox, Crotalus cerastes, Crotalus viridis and Sistrurus catenatus. This pattern implies that the rattling "could serve as a general attention-getting device", which "is designed as a deimatic or startle display". Its similarity to the "broadband, harsh sounds" used as warning calls by birds and mammals may enhance its effectiveness. Since rattlesnakes can barely hear the sound, it is unlikely to serve as any form of communication to other snakes of the same species. Finally, the sounds are not in themselves loud enough to cause pain and hence keep predators away.

Fenton and Licht note that the effect of a rattlesnake's rattling could be deimatic (startle) in inexperienced animals, whether predators or large animals that might injure the snake by stepping on it, but aposematic (a warning signal) in animals that are aware of the rattle's meaning. They refer to the work of Fenton and his colleague David Bates on the responses of the big brown bat, Eptesicus fuscus, to the defensive clicks made by moths in the family Arctiidae, which includes the garden tiger moth, Arctia caja. This family includes large, furry, bitter-tasting or poisonous moths. They found that while sounds can startle inexperienced bats, after a few trials the bats ignored the sounds if the prey was edible; but the same sounds can warn experienced bats of bitter-tasting prey (an honest signal).

Rattlesnake raising rattle on tail, drawn by St. George Mivart, on the Genesis of Species. The rattle may both startle inexperienced predators and warn off experienced ones

Garden tiger moth, Arctia caja, displays startling bright pattern of black spots on orange-red hindwings. The insect is bitter-tasting, so the pattern may be aposematic as well as deimatic

Collective Animal Behavior

Collective animal behavior is a form of social behavior involving the coordinated behavior of large groups of similar animals as well as emergent properties of these groups. This can include the costs and benefits of group membership, the transfer of information across the group, the group decision-making process, and group locomotion and synchronization. Studying the principles of collective animal behavior has relevance to human engineering problems through the philosophy of biomimetics. For instance, determining the rules by which an individual animal navigates relative to its neighbors in a group can lead to advances in the deployment and control of groups of swimming or flying micro-robots such as UAVs (Unmanned Aerial Vehicles).

Starling flock asunset in Denmark

Examples

Examples of collective animal behavior include:

- Flocking birds
- Herding ungulates
- Shoaling and schooling fish
- Swarming Antarctic krill
- Pods of dolphins
- Marching locusts
- Nest building ants

Proposed Functions

Many functions of animal aggregations have been proposed. These proposed functions may be grouped into the four following categories: social and genetic, anti-predator, enhanced foraging, and increased locomotion efficiency.

Social Interaction

Support for the social and genetic function of aggregations, especially those formed by fish, can be seen in several aspects of their behavior. For instance, experiments have shown that individual fish removed from a school will have a higher respiratory rate than those found in the school. This effect has been partly attributed to stress, although hydrodynamic factors were considered more important in this particular study. The calming effect of being with conspecifics may thus provide a social motivation for remaining in an aggregation. Herring, for instance, will become very agitated if they are isolated from conspecifics. Fish schools have also been proposed to serve a reproductive function since they provide increased access to potential mates.

Protection from Predators

School of goldband fusiliers

Several anti-predator functions of animal aggregations have been proposed. One potential method by which fish schools or bird flocks may thwart predators is the 'predator confusion effect' proposed and demonstrated by Milinski and Heller (1978). This theory is based on the idea that it becomes difficult for predators to pick out individual prey from groups because the many moving targets create a sensory overload of the predator's visual channel. Milinski and Heller's findings have been corroborated both in experiment and computer simulations.

A second potential anti-predator effect of animal aggregations is the "many eyes" hypothesis. This theory states that as the size of the group increases, the task of scanning the environment for predators can be spread out over many individuals. Not only does this mass collaboration presumably provide a higher level of vigilance, it could also allow more time for individual feeding.

A third hypothesis for an anti-predatory effect of animal aggregation is the "encounter dilution" effect. Hamilton, for instance, proposed that the aggregation of animals was due to a "selfish" avoidance of a predator and was thus a form of cover-seeking. Another formulation of the theory was given by Turner and Pitcher and was viewed as a combination of detection and attack probabilities. In the detection component of the theory, it was suggested that potential prey might benefit by living together since a predator is less likely to chance upon a single group than a scattered distribution. In the attack component, it was thought that an attacking predator is less likely to eat a particular animal when a greater number of individuals are present. In sum, an individual has an advantage if it is in the larger of two groups, assuming that the probability of detection and attack does not increase disproportionately with the size of the group.

Enhanced Foraging

A third proposed benefit of animal groups is that of enhanced foraging. This ability was demonstrated by Pitcher and others in their study of foraging behavior in shoaling cyprinids. In this study, the time it took for groups of minnows and goldfish to find a patch of food was quantified. The number of fishes in the groups was varied, and a statistically significant decrease in the amount of time necessary for larger groups to find food was established. Further support for an enhanced foraging capability of schools is seen in the structure of schools of predatory fish. Partridge and others analyzed the school structure of Atlantic bluefin tuna from aerial photographs and found that the school assumed a parabolic shape, a fact that was suggestive of cooperative hunting in this species.

Increased Locomotion Efficiency

This theory states that groups of animals moving in a fluid environment may save energy when swimming or flying together, much in the way that bicyclists may draft one another in a peloton. Geese flying in a Vee formation are also thought to save energy by flying in the updraft of the wing-tip vortex generated by the previous animal in the formation. Ducklings have also been shown to save energy by swimming in a line. Increased efficiencies in swimming in groups have also been proposed for schools of fish and Antarctic krill.

Group Structure

The structure of large animal groups has been difficult to study because of the large number of animals involved. The experimental approach is therefore often complemented by mathematical modeling of animal aggregations.

Experimental Approach

Experiments investigating the structure of animal aggregations seek to determine the 3D position of each animal within a volume at each point in time. It is important to know the internal structure of the group because that structure can be related to the proposed motivations for animal grouping. This capability requires the use of multiple cameras trained on the same volume in space, a technique known as stereophotogrammetry. When hundreds or thousands of animals occupy the study volume, it becomes difficult to identify each one. In addition, animals may block one another in the camera views, a problem known as occlusion. Once the location of each animal at each point in time is known, various parameters describing the animal group can be extracted.

These parameters include:

1. *Density*: The density of an animal aggregation is the number of animals divided by the volume (or area) occupied by the aggregation. Density may not be a constant throughout the group. For instance, starling flocks have been shown to maintain higher densities on the edges than in the middle of the flock, a feature that is presumably related to defense from predators.

2. *Polarity*: The group polarity describes if the group animals are all pointing in the same direction or not. In order to determine this parameter, the average orientation of all animals in the group is determined. For each animal, the angular difference between its orientation and the group orientation is then found. The group polarity is then the average of these differences.

3. *Nearest Neighbor Distance*: The nearest neighbor distance (NND) describes the distance between the centroid of one animal (the focal animal) and the centroid of the animal nearest to the focal animal. This parameter can be found for each animal in an aggregation and then averaged. Care must be taken to account for the animals located at the edge of an animal aggregation. These animals have no neighbor in one direction.

4. *Nearest Neighbor Position*: In a polar coordinate system, the nearest neighbor position describes the angle and distance of the nearest neighbor to a focal animal.

5. *Packing Fraction*: Packing fraction is a parameter borrowed from physics to define the organization (or state i.e. solid, liquid, or gas) of 3D animal groups. It is an alternative measure to density. In this parameter, the aggregation is idealized as an ensemble of solid spheres, with each animal at the center of a sphere. The packing fraction is defined as the ratio of the total volume occupied by all individual spheres divided by the global volume of the aggregation. Values range from zero to one, where a small packing fraction represents a dilute system like a gas. Cavagna found that the packing fraction for groups of starlings was 0.012.

6. *Integrated Conditional Density*: This parameter measures the density at various length scales and therefore describes the homogeneity of density throughout an animal group.

7. *Pair Distribution Function*: This parameter is usually used in physics to characterize the degree of spatial order in a system of particles. It also describes the density, but this measures describes the density at a distance away from a given point. Cavagna et al. found that flocks of starlings exhibited more structure than a gas but less than a liquid.

Modeling Approach

The simplest mathematical models of animal aggregations generally instruct the individual animals to follow three rules:

1. Move in the same direction as your neighbor
2. Remain close to your neighbors
3. Avoid collisions with your neighbors

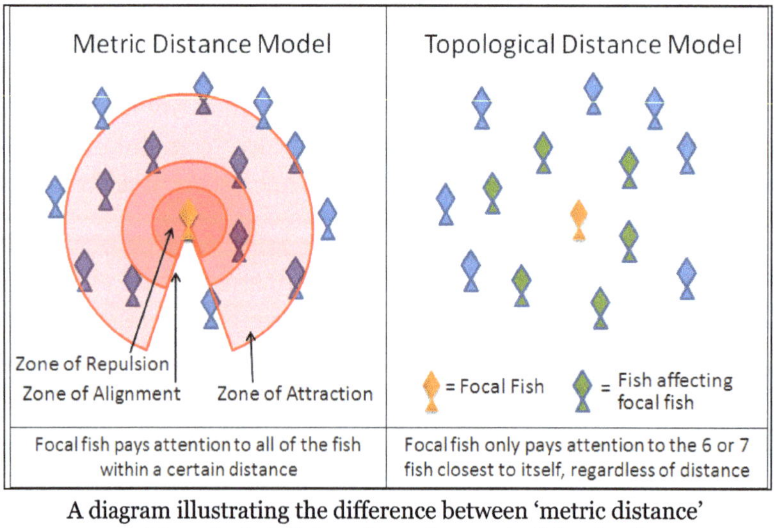

A diagram illustrating the difference between 'metric distance' and 'topological distance' in reference to fish schools

An example of such a simulation is the Boids program created by Craig Reynolds in 1986. Another is the Self Propelled Particle model. Many current models use variations on these rules. For instance, many models implement these three rules through layered zones around each animal. In the zone of repulsion very close to the animal, the focal animal will seek to distance itself from its neighbors in order to avoid a collision. In the slightly further away zone of alignment, a focal animal will seek to align its direction of motion with its neighbors. In the outmost zone of attraction, which extends as far away from the focal animal as it is able to sense, the focal animal will seeks to move towards a neighbor. The shape of these zones will necessarily be affected by the sensory capabilities of the animal. For example, the visual field of a bird does not extend behind its body. Fish, on the other hand, rely on both vision and on hydrodynamic signals relayed through its lateral line. Antarctic krill rely on vision and on hydrodynamic signals relayed through its antennae.

Recent studies of starling flocks have shown, however, that each bird modifies its position relative to the six or seven animals directly surrounding it, no matter how close or how far away those animals are. Interactions between flocking starlings are thus based on a topological rule rather than a metric rule. It remains to be seen whether the same rule can be applied to other animals. Another recent study, based on an analysis of high speed camera footage of flocks above Rome and assuming minimal behavioural rules, has convincingly simulated a number of aspects of flock behaviour.

Collective Decision Making

Aggregations of animals are faced with decisions which they must make if they are to remain together. For a school of fish, an example of a typical decision might be which direction to swim when confronted by a predator. Social insects such as ants and bees must collectively decide where to build a new nest. A herd of elephants must decide when and where to migrate. How are these decisions made? Do stronger or more experienced 'leaders' exert more influence than other group members, or does the group make a decision by consensus? The answer probably depends on the species. While the role of a leading matriarch in an elephant herd is well known, studies have shown that some animal species use a consensus approach in their collective decision-making process.

A recent investigation showed that small groups of fish used consensus decision-making when deciding which fish model to follow. The fish did this by a simple quorum rule such that individuals watched the decisions of others before making their own decisions. This technique generally resulted in the 'correct' decision but occasionally cascaded into the 'incorrect' decision. In addition, as the group size increased, the fish made more accurate decisions in following the more attractive fish model. Consensus decision-making, a form of collective intelligence, thus effectively uses information from multiple sources to generally reach the correct conclusion.

Some simulations of collective decision-making use the Condorcet method to model the way groups of animals come to consensus.

Animal Consciousness

Animal consciousness, or animal awareness, is the quality or state of self-awareness within an animal, or of being aware of an external object or something within itself. In humans, consciousness has been defined as: sentience, awareness, subjectivity, qualia, the ability to experience or to feel, wakefulness, having a sense of selfhood, and the executive control system of the mind. Despite the difficulty in definition, many philosophers believe there is a broadly shared underlying intuition about what consciousness is.

The topic of animal consciousness is beset with a number of difficulties. It poses the problem of other minds in an especially severe form because animals, lacking the ability to use human language, cannot tell us about their experiences. Also, it is difficult to reason objectively about the question, because a denial that an animal is conscious is often taken to imply that it does not feel, its life has no value, and that harming it is not morally wrong. The 17th-century French philosopher René Descartes, for example, has sometimes been blamed for mistreatment of animals because he argued that only humans are conscious.

According to the Cambridge Declaration on Consciousness, "near human-like levels of consciousness" have been observed in the grey parrot

Philosophers who consider subjective experience the essence of consciousness also generally believe, as a correlate, that the existence and nature of animal consciousness can never rigorously be known. The American philosopher Thomas Nagel spelled out this point of view in an influential essay titled *What Is it Like to Be a Bat?*. He said that an organism is conscious "if and only if there is something that it is like to be that organism—something it is like *for* the organism"; and he argued that no matter how much we know about an animal's brain and behavior, we can never really put ourselves into the mind of the animal and experience its world in the way it does itself. Other thinkers, such as the cognitive scientist Douglas Hofstadter, dismiss this argument as incoherent. Several psychologists and ethologists have argued for the existence of animal consciousness by describing a range of behaviors that appear to show animals holding beliefs about things they cannot directly perceive—Donald Griffin's 2001 book *Animal Minds* reviews a substantial portion of the evidence.

Animal consciousness has been actively researched for over one hundred years. In 1927 the American functional psychologist Harvey Carr argued that any valid measure or understanding of awareness in animals depends on "an accurate and complete knowledge of its essential conditions in man". A more recent review concluded in 1985 that "the best approach is to use experiment (especially psychophysics) and observation to trace the dawning and ontogeny of self-consciousness, perception, communication, intention, beliefs, and reflection in normal human fetuses, infants, and children". In 2012, a group of neuroscientists signed the Cambridge Declaration on Consciousness, which "unequivocally" asserted that "humans are not unique in possessing the neurological substrates that generate consciousness. Non-human animals, including all mammals and birds, and many other creatures, including octopuses, also possess these neural substrates."

Philosophical Background

René Descartes argued that only humans are conscious, and not other animals

The mind–body problem in philosophy examines the relationship between mind and matter, and in particular the relationship between consciousness and the brain. A variety of approaches have been proposed. Most are either dualist or monist. Dualism maintains a rigid distinction between the realms of mind and matter. Monism maintains that there is only one kind of stuff, and that mind and matter are both aspects of it. The problem was addressed by pre-Aristotelian philosophers, and was famously addressed by René Descartes in the 17th century, resulting in Cartesian dualism. Descartes believed that humans only, and not other animals have this non-physical mind.

The rejection of the mind–body dichotomy is found in French Structuralism, and is a position that generally characterized post-war French philosophy. The absence of an empirically identifiable meeting point between the non-physical mind and its physical extension has proven problematic to dualism and many modern philosophers of mind maintain that the mind is not something separate from the body. These approaches have been particularly influential in the sciences, particularly in the fields of sociobiology, computer science, evolutionary psychology, and the neurosciences.

Epiphenomenalism

Epiphenomenalism is the theory in philosophy of mind that mental phenomena are caused by physical processes in the brain or that both are effects of a common cause, as opposed to mental phenomena driving the physical mechanics of the brain. The impression that thoughts, feelings, or sensations cause physical effects, is therefore to be understood as illusory to some extent. For example, it is not the feeling of fear that produces an increase in heart beat, both are symptomatic of a common physiological origin, possibly in response to a legitimate external threat.

The history of epiphenomenalism goes back to the post-Cartesian attempt to solve the riddle of Cartesian dualism, i.e., of how mind and body could interact. La Mettrie, Leibniz and Spinoza all in their own way began this way of thinking. The idea that even if the animal were conscious nothing would be added to the production of behavior, even in animals of the human type, was first voiced by La Mettrie (1745), and then by Cabanis (1802), and was further explicated by Hodgson (1870) and Huxley (1874). Huxley (1874) likened mental phenomena to the whistle on a steam locomotive. However, epiphenomenalism flourished primarily as it found a niche among methodological or scientific behaviorism. In the early 1900s scientific behaviorists such as Ivan Pavlov, John B. Watson, and B. F. Skinner began the attempt to uncover laws describing the relationship between stimuli and responses, without reference to inner mental phenomena. Instead of adopting a form of eliminativism or mental fictionalism, positions that deny that inner mental phenomena exist, a behaviorist was able to adopt epiphenomenalism in order to allow for the existence of mind. However, by the 1960s, scientific behaviourism met substantial difficulties and eventually gave way to the cognitive revolution. Participants in that revolution, such as Jerry Fodor, reject epiphenomenalism and insist upon the efficacy of the mind. Fodor even speaks of "epiphobia"—fear that one is becoming an epiphenomenalist.

Thomas Henry Huxley defends in an essay titled *On the Hypothesis that Animals are Automata, and its History* an epiphenomenalist theory of consciousness according to which consciousness is a causally inert effect of neural activity—"as the steam-whistle which accompanies the work of a locomotive engine is without influence upon its machinery". To this William James objects in his essay *Are We Automata?* by stating an evolutionary argument for mind-brain interaction implying that if the preservation and development of consciousness in the biological evolution is a result of natural selection, it is plausible that consciousness has not only been influenced by neural processes, but has had a survival value itself; and it could only have had this if it had been efficacious. Karl Popper develops in the book *The Self and Its Brain* a similar evolutionary argument.

Animal Ethics

Bernard Rollin of Colorado State University, the principal author of two U.S. federal laws regulating pain relief for animals, writes that researchers remained unsure into the 1980s as to whether animals experience pain, and veterinarians trained in the U.S. before 1989 were simply taught to ignore animal pain. In his interactions with scientists and other veterinarians, Rollin was regularly asked to prove animals are conscious and provide scientifically acceptable grounds for claiming they feel pain. Academic reviews of the topic are equivocal, noting that the argument that animals have at least simple conscious thoughts and feelings has strong support, but some critics continue to question how reliably animal mental states can be determined. A refereed journal *Animal Sentience* launched in 2015 by the Institute of Science and Policy of The Humane Society of the United States is devoted to research on this and related topics.

Defining Consciousness

Consciousness is an elusive concept that presents many difficulties when attempts are made to define it. Its study has progressively become an interdisciplinary challenge for numerous researchers, including ethologists, neurologists, cognitive neuroscientists, philosophers, psychologists and psychiatrists.

> "About forty meanings attributed to the term *consciousness* can be identified and categorized based on *functions* and *experiences*. The prospects for reaching any single, agreed-upon, theory-independent definition of consciousness appear remote."

In 1976 Richard Dawkins wrote, "The evolution of the capacity to simulate seems to have culminated in subjective consciousness. Why this should have happened is, to me, the most profound mystery facing modern biology". In 2004, eight neuroscientists felt it was still too soon for a definition. They wrote an apology in "Human Brain Function":

> "We have no idea how consciousness emerges from the physical activity of the brain and we do not know whether consciousness can emerge from non-biological systems, such as computers. At this point the reader will expect to find a careful and precise definition of consciousness. You will be disappointed. Consciousness has not yet become a scientific term that can be defined in this way. Currently we all use the term *consciousness* in many different and often ambiguous ways. Precise definitions of different aspects of consciousness will emerge but to make precise definitions at this stage is premature."

Consciousness is sometimes defined as the quality or state of being aware of an external object or something within oneself. It has been defined somewhat vaguely as: subjectivity, awareness, sentience, the ability to experience or to feel, wakefulness, having a sense of selfhood, and the executive control system of the mind. Despite the difficulty in definition, many philosophers believe that there is a broadly shared underlying intuition about what consciousness is. Max Velmans and Susan Schneider wrote in *The Blackwell Companion to Consciousness*: "Anything that we are aware of at a given moment forms part of our consciousness, making conscious experience at once the most familiar and most mysterious aspect of our lives."

Related terms, also often used in vague or ambiguous ways, are:

- Awareness: the state or ability to perceive, to feel, or to be conscious of events, objects, or sensory patterns. In this level of consciousness, sense data can be confirmed by an observer without necessarily implying understanding. More broadly, it is the state or quality of being aware of something. In biological psychology, awareness is defined as a human's or an animal's perception and cognitive reaction to a condition or event.
- Self-awareness: the capacity for introspection and the ability to reconcile oneself as an individual separate from the environment and other individuals.
- Self-consciousness: an acute sense of self-awareness. It is a preoccupation with oneself, as opposed to the philosophical state of self-awareness, which is the awareness that one exists as an individual being; although some writers use both terms interchangeably or synonymously.
- Sentience: the ability to be aware (feel, perceive, or be conscious) of one's surroundings or to have subjective experiences. Sentience is a minimalistic way of defining consciousness, which is otherwise commonly used to collectively describe sentience plus other characteristics of the mind.
- Sapience: often defined as wisdom, or the ability of an organism or entity to act with appropriate judgment, a mental faculty which is a component of intelligence or alternatively may be considered an additional faculty, apart from intelligence, with its own properties.
- Qualia: individual instances of subjective, conscious experience.

Sentience (the ability to feel, perceive, or to experience subjectivity) is not the same as self-awareness (being aware of oneself as an individual). The mirror test is sometimes considered to be an operational test for self-awareness, and the handful of animals that have passed it are often considered to be self-aware. It remains debatable whether recognition of one's mirror image can be properly construed to imply full self-awareness, particularly given that robots are being constructed which appear to pass the test.

Much has been learned in neuroscience about correlations between brain activity and subjective, conscious experiences, and many suggest that neuroscience will ultimately explain consciousness; "consciousness is a biological process that will eventually be explained in terms of molecular signaling pathways used by interacting populations of nerve cells". However, this view has been criticized because consciousness has yet to be shown to be a process, and the so-called "hard problem" of relating consciousness directly to brain activity remains elusive.

Scientific Approaches

Since Descartes's proposal of dualism, it became a general consensus that the mind had become a matter of philosophy and that science was not able to penetrate the issue of consciousness - that consciousness was outside of space and time. However, over the last 20 years, many scholars have begun to move toward a science of consciousness. Antonio Damasio and Gerald Edelman are two neuroscientists who have led the move to neural correlates of the self and of consciousness. Damasio has demonstrated that emotions and their biological foundation play a critical role in high level cognition, and Edelman has created a framework for analyzing consciousness through a scientific outlook. The current problem consciousness researchers face involves explaining how and why consciousness arises from neural computation. In his research on this problem, Edelman has developed a theory of consciousness, in which he has coined the terms primary consciousness and secondary consciousness.

Eugene Linden, author of *The Parrot's Lament* suggests there are many examples of animal behavior and intelligence that surpass what people would suppose to be the boundary of animal consciousness. Linden contends that in many of these documented examples, a variety of animal species exhibit behavior that can only be attributed to emotion, and to a level of consciousness that we would normally ascribe only to our own species.

Philosopher Daniel Dennett counters that:

Consciousness requires a certain kind of informational organization that does not seem to be 'hard-wired' in humans, but is instilled by human culture. Moreover, consciousness is not a black-or-white, all-or-nothing type of phenomenon, as is often assumed. The differences between humans and other species are so great that speculations about animal consciousness seem ungrounded. Many authors simply assume that an animal like a bat has a point of view, but there seems to be little interest in exploring the details involved.

Consciousness in mammals (including humans) is an aspect of the mind generally thought to comprise qualities such as subjectivity, sentience, and the ability to perceive the relationship between oneself and one's environment. It is a subject of much research in philosophy of mind, psychology, neuroscience, and cognitive science. Some philosophers divide consciousness into phenomenal consciousness, which is subjective experience itself, and access consciousness, which refers to the global availability of information to processing systems in the brain. Phenomenal consciousness has many different experienced qualities, often referred to as qualia. Phenomenal consciousness is usually consciousness *of* something or *about* something, a property known as intentionality in philosophy of mind.

In humans, there are three common methods of studying consciousness, i.e. verbal report, behavioural demonstrations, and neural correlation with conscious activity. Unfortunately these can only be generalized to non-human taxa with varying degrees of difficulty.

Mirror Test

Elephants can recognize themselves in a mirror

The sense in which animals (or human infants) can be said to have consciousness or a self-concept has been hotly debated; it is often referred to as the debate over animal minds. The best known research technique in this area is the mirror test devised by Gordon G. Gallup, in which the skin of an animal (or human infant) is marked, while it is asleep or sedated, with a mark that cannot be seen directly but is visible in a mirror. The animal is then allowed to see its reflection in a mirror; if the animal spontaneously directs grooming behaviour towards the mark, that is taken as an indication that it is aware of itself. Over the past 30 years, many studies have found evidence that animals recognise themselves in mirrors. Self-awareness by this criterion has been reported for:

Apes

- Common chimpanzees, bonobos, orangutans and gorillas.

Other land mammals

- Elephants.

Cetaccans

- Bottlenose dolphin (*Tursiops truncatus*).
- Killer whales: Killer whales (*Orcinus orca*) and false killer whales (*Pseudorca crassidens*) may be able to recognize themselves in mirrors.

Birds

- Magpies.

Until recently it was thought that self-recognition was absent from animals without a neocortex, and was restricted to mammals with large brains and well developed social cognition. However, in 2008 a study of self-recognition in corvids reported significant results for magpies. Mammals and birds inherited the same brain components from their last common ancestor nearly 300 million years ago, and have since independently evolved and formed significantly different brain types. The results of the mirror and mark tests showed that neocortex-less magpies are capable of understanding that a mirror image belongs to their own body. The findings show that magpies respond in the mirror and mark test in a manner similar to apes, dolphins and elephants. This is a remarkable capability that, although not fully concrete in its determination of self-recognition, is at least a prerequisite of self-recognition. This is not only of interest regarding the convergent evolution of social intelligence; it is also valuable for an understanding of the general principles that govern cognitive evolution and their underlying neural mechanisms. The magpies were chosen to study based on their empathy/lifestyle, a possible precursor for their ability of self-awareness. However even in the chimpanzee, the species most studied and with the most convincing findings, clear-cut evidence of self-recognition is not obtained in all individuals tested. Occurrence is about 75% in young adults and considerably less in young and old individuals. For monkeys, non-primate mammals, and in a number of bird species, exploration of the mirror and social displays were observed. However, hints at mirror-induced self-directed behavior have been obtained.

The mirror test has attracted controversy among some researchers because it is entirely focused on vision, the primary sense in humans, while other species rely more heavily on other senses such as the olfactory sense in dogs. A study in 2015 showed that the "sniff test of self-recognition (STSR)" provides evidence of self-awareness in dogs.

Language

Another approach to determine whether a non-human animal is conscious derives from passive speech research with a macaw (see Arielle). Some researchers propose that by passively listening to an animal's voluntary speech, it is possible to learn about the thoughts of another creature and to determine that the speaker is conscious. This type of research was originally used to investigate a child's crib speech by Weir (1962) and in investigations of early speech in children by Greenfield and others (1976).

Zipf's law might be able to be used to indicate if a given dataset of animal communication indicate an intelligent natural language. Some researchers have used this algorithm to study bottlenose dolphin language.

Pain or Suffering

Further arguments revolve around the ability of animals to feel pain or suffering. Suffering implies consciousness. If animals can be shown to suffer in a way similar or identical to humans, many of the arguments against human suffering could then, presumably, be extended to animals. Others have argued that pain can be demonstrated by adverse reactions to negative stimuli that are non-purposeful or even maladaptive. One such reaction is transmarginal inhibition, a phenomenon observed in humans and some animals akin to mental breakdown.

Carl Sagan, the American cosmologist, points to reasons why humans have had a tendency to deny animals can suffer:

Humans – who enslave, castrate, experiment on, and fillet other animals – have had an understandable penchant for pretending animals do not feel pain. A sharp distinction between humans and 'animals' is essential if we are to bend them to our will, make them work for us, wear them, eat them – without any disquieting tinges of guilt or regret. It is unseemly of us, who often behave so unfeelingly toward other animals, to contend that only humans can suffer. The behavior of other animals renders such pretensions specious. They are just too much like us.

John Webster, a professor of animal husbandry at Bristol, argues:

People have assumed that intelligence is linked to the ability to suffer and that because animals have smaller brains they suffer less than humans. That is a pathetic piece of logic, sentient animals have the capacity to experience pleasure and are motivated to seek it, you only have to watch how cows and lambs both seek and enjoy pleasure when they lie with their heads raised to the sun on a perfect English summer's day. Just like humans.

However, there is no agreement where the line between organisms that can feel pain and those that cannot should be drawn. Justin Leiber, a philosophy professor at Oxford University writes that:

Montaigne is ecumenical in this respect, claiming consciousness for spiders and ants, and even writing of our duties to trees and plants. Singer and Clarke agree in denying consciousness to sponges. Singer locates the distinction somewhere between the shrimp and the oyster. He, with rather considerable convenience for one who is thundering hard accusations at others, slides by the case of insects and spiders and bacteria, they pace Montaigne, apparently and rather conveniently do not feel pain. The intrepid Midgley, on the other hand, seems willing to speculate about the subjective experience of tapeworms Nagel appears to draw the line at flounders and wasps, though more recently he speaks of the inner life of cockroaches.

There are also some who reject the argument entirely, arguing that although suffering animals feel anguish, a suffering plant also struggles to stay alive (albeit in a less visible way). In fact, no living organism 'wants' to die for another organism's sustenance. In an article written for the New York Times, Carol Kaesuk Yoon argues that:

When a plant is wounded, its body immediately kicks into protection mode. It releases a bouquet of volatile chemicals, which in some cases have been shown to induce neighboring plants to pre-emptively step up their own chemical defenses and in other cases to lure in predators of the beasts that may be causing the damage to the plants. Inside the plant, repair systems are engaged and defenses are mounted, the molecular details of which scientists are still working out, but which involve signaling molecules coursing through the body to rally the cellular troops, even the enlisting of the genome itself, which begins churning out defense-related proteins ... If you think about it, though, why would we expect any organism to lie down and die for our dinner? Organisms have evolved to do everything in their power to avoid being extinguished. How long would any lineage be likely to last if its members effectively didn't care if you killed them?

Cognitive Bias and Emotion

Is the glass half empty or half full?

Cognitive bias in animals is a pattern of deviation in judgment, whereby inferences about other animals and situations may be drawn in an illogical fashion. Individuals create their own "subjective social reality" from their perception of the input. It refers to the question "Is the glass half empty or half full?", used as an indicator of optimism or pessimism. Cognitive biases have been shown in a wide range of species including rats, dogs, rhesus macaques, sheep, chicks, starlings and honeybees.

The neuroscientist Joseph LeDoux advocates avoiding terms derived from human subjective experience when discussing brain functions in animals. For example, the common practice of calling brain circuits that detect and respond to threats "fear circuits" implies that these circuits are responsible for feelings of fear. LeDoux argues that Pavlovian fear conditioning should be renamed Pavlovian threat conditioning to avoid the implication that "fear" is being acquired in rats or humans. Key to his theoretical change is the notion of survival functions mediated by survival circuits, the purpose of which is to keep organisms alive rather than to make emotions. For example, defensive survival circuits exist to detect and respond to threats. While all organisms can do this, only organisms that can be conscious of their own brain's activities can feel fear. Fear is a conscious experience and occurs the same way as any other kind of conscious experience: via cortical circuits that allow attention to certain forms of brain activity. LeDoux argues the only differences between an emotional and non-emotion state of consciousness are the underlying neural ingredients that contribute to the state.

Neuroscience

Neuroscience is the scientific study of the nervous system. It is a highly active interdisciplinary science that collaborates with many other fields. The scope of neuroscience has broadened recently to include molecular, cellular, developmental, structural, functional, evolutionary, computational, and medical aspects of the nervous system. Theoretical studies of neural networks are being complemented with techniques for imaging sensory and motor tasks in the brain. According to a 2008 paper, neuroscience explanations of psychological phenomena currently have a "seductive allure", and "seem to generate more public interest" than explanations which do not contain neuroscientific information. They found that subjects who were not neuroscience experts "judged that explanations with logically irrelevant neuroscience information were more satisfying than explanations without.

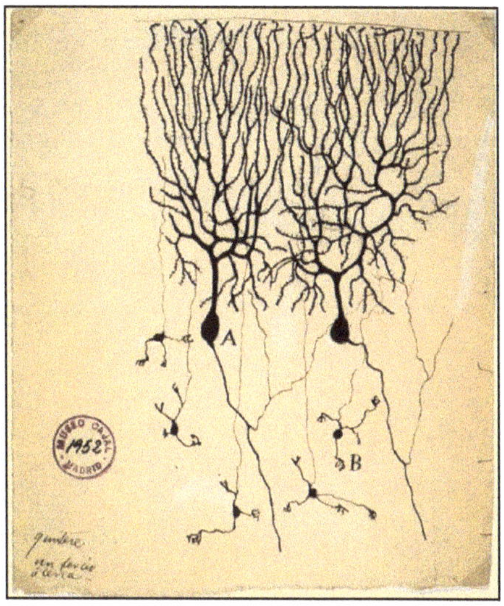

Drawing by Santiago Ramón y Cajal of neurons in the pigeon cerebellum

Neural Correlates

The neural correlates of consciousness constitute the minimal set of neuronal events and mechanisms sufficient for a specific conscious percept. Neuroscientists use empirical approaches to discover neural correlates of subjective phenomena. The set should be *minimal* because, if the brain is sufficient to give rise to any given conscious experience, the question is which of its components is necessary to produce it.

Visual sense and representation was reviewed in 1998 by Francis Crick and Christof Koch. They concluded sensory neuroscience can be used as a bottom-up approach to studying consciousness, and suggested experiments to test various hypotheses in this research stream.

A feature that distinguishes humans from most animals is that we are not born with an extensive repertoire of behavioral programs that would enable us to survive on our own ("physiological prematurity"). To compensate for this, we have an unmatched ability to learn, i.e., to consciously acquire such programs by imitation or exploration. Once consciously acquired and sufficiently exercised, these programs can become automated to the extent that their execution happens beyond the realms of our awareness. Take, as an example, the incredible fine motor skills exerted in playing a Beethoven piano sonata or the sensorimotor coordination required to ride a motorcycle along a curvy mountain road. Such complex behaviors are possible only because a sufficient number of the subprograms involved can be executed with minimal or even suspended conscious control. In fact, the conscious system may actually interfere somewhat with these automated programs.

The growing ability of neuroscientists to manipulate neurons using methods from molecular biology in combination with optical tools depends on the simultaneous development of appropriate behavioural assays and model organisms amenable to large-scale genomic analysis and manipulation. A combination of such fine-grained neuronal analysis in animals with ever more sensitive psychophysical and brain imaging techniques in humans, complemented by the development of a robust theoretical predictive framework, will hopefully lead to a rational understanding of consciousness.

Neocortex

Previously researchers had thought that patterns of neural sleep exhibited by zebra finches needed a mammalian neocortex

The neocortex is a part of the brain of mammals. It consists of the grey matter, or neuronal cell bodies and unmyelinated fibers, surrounding the deeper white matter (myelinated axons) in the cerebrum. The neocortex is smooth in rodents and other small mammals, whereas in primates and other larger mammals it has deep grooves and wrinkles. These folds increase the surface area of the neocortex considerably without taking up too much more volume. Also, neurons within the same wrinkle have more opportunity for connectivity, while neurons in different wrinkles have less opportunity for connectivity, leading to compartmentalization of the cortex. The neocortex is divided into frontal, parietal, occipital, and temporal lobes, which perform different functions. For example, the occipital lobe contains the primary visual cortex, and the temporal lobe contains the primary auditory cortex. Further subdivisions or areas of neocortex are responsible for more specific cognitive processes. The neocortex is the newest part of the cerebral cortex to evolve (hence the prefix "neo"); the other parts of the cerebral cortex are the paleocortex and archicortex, collectively known as the allocortex. In humans, 90% of the cerebral cortex is neocortex.

Researchers have argued that consciousness arises in the neocortex, and therefore cannot arise in animals which lack a neocortex. For example, Rose argued in 2002 that the "fishes have nervous systems that mediate effective escape and avoidance responses to noxious stimuli, but, these responses must occur without a concurrent, human-like awareness of pain, suffering or distress, which depend on separately evolved neocortex." Recently that view has been challenged, and many researchers now believe that animal consciousness can arise from homologous subcortical brain networks.

Attention

Attention is the cognitive process of selectively concentrating on one aspect of the environment while ignoring other things. Attention has also been referred to as the allocation of processing resources. Attention also has variations amongst cultures. Voluntary attention develops in specific cultural and institutional contexts through engagement in cultural activities with more competent community members.

Most experiments show that one neural correlate of attention is enhanced firing. If a neuron has a certain response to a stimulus when the animal is not attending to the stimulus, then when the animal does attend to the stimulus, the neuron's response will be enhanced even if the physical characteristics of the stimulus remain the same. In many cases attention produces changes in the EEG. Many animals, including humans, produce gamma waves (40–60 Hz) when focusing attention on a particular object or activity.

Extended Consciousness

Extended consciousness is an animal's autobiographical self-perception. It is thought to arise in the brains of animals which have a substantial capacity for memory and reason. It does not necessarily require language. The perception of a historic and future self arises from a stream of information from the immediate environment and from neural structures related to memory. The concept was popularised by Antonio Damasio and is used in biological psychology. Extended consciousness is said to arise in structures in the human brain described as *image spaces* and *dispositional spaces*. Image spaces imply areas where sensory impressions of all types are processed, including the focused awareness of the core consciousness. Dispositional spaces include convergence zones, which are networks in the brain where memories are processed and recalled, and where knowledge is merged with immediate experience.

Metacognition

Metacognition is defined as "cognition about cognition", or "knowing about knowing." It can take many forms; it includes knowledge about when and how to use particular strategies for learning or for problem solving. It has been suggested that metacognition in some animals provides evidence for cognitive self-awareness. There are generally two components of metacognition: knowledge about cognition, and regulation of cognition. Writings on metacognition can be traced back at least as far as *De Anima* and the *Parva Naturalia* of the Greek philosopher Aristotle. Metacognologists believe that the ability to consciously think about thinking is unique to sapient species and indeed is one of the definitions of sapience. There is evidence that rhesus monkeys and apes can make accurate judgments about the strengths of their memories of fact and monitor their own uncertainty, while attempts to demonstrate metacognition in birds have been inconclusive. A 2007 study provided some evidence for metacognition in rats, but further analysis suggested that they may have been following simple operant conditioning principles, or a behavioral economic model.

Mirror Neurons

Mirror neurons are neurons that fire both when an animal acts and when the animal observes the same action performed by another. Thus, the neuron "mirrors" the behavior of the other, as though the observer were itself acting. Such neurons have been directly observed in primate and other species including birds. The function of the mirror system is a subject of much speculation. Many researchers in cognitive neuroscience and cognitive psychology consider that this system provides the physiological mechanism for the perception action coupling. They argue that mirror neurons may be important for understanding the actions of other people, and for learning new skills by imitation. Some researchers also speculate that mirror systems may simulate observed actions, and thus contribute to theory of mind skills, while others relate mirror neurons to language abilities. Neuroscientists such as Marco Iacoboni (UCLA) have argued that mirror neuron systems in the human brain help us understand the actions and intentions of other people. In a study published in March 2005 Iacoboni and his colleagues reported that mirror neurons could discern if another person who was picking up a cup of tea planned to drink from it or clear it from the table. In addition, Iacoboni and a number of other researchers have argued that mirror neurons are the neural basis of the human capacity for emotions such as empathy. Vilayanur S. Ramachandran has speculated that mirror neurons may provide the neurological basis of self-awareness.

Evolutionary Psychology

Consciousness is likely an evolved adaptation since it meets George Williams' criteria of species universality, complexity, and functionality, and it is a trait that apparently increases fitness. Opinions are divided as to where in biological evolution consciousness emerged and about whether or not consciousness has survival value. It has been argued that consciousness emerged (i) exclusively with the first humans, (ii) exclusively with the first mammals, (iii) independently in mammals and birds, or (iv) with the first reptiles. Donald Griffin suggests in his book *Animal Minds* a gradual evolution of consciousness. Each of these scenarios raises the question of the possible survival value of consciousness.

In his paper "Evolution of consciousness," John Eccles argues that special anatomical and physical adaptations of the mammalian cerebral cortex gave rise to consciousness. In contrast, others have argued that the recursive circuitry underwriting consciousness is much more primitive, having evolved initially in pre-mammalian species because it improves the capacity for interaction with both social *and* natural environments by providing an energy-saving "neutral" gear in an otherwise energy-expensive motor output machine. Once in place, this recursive circuitry may well have provided a basis for the subsequent development of many of the functions that consciousness facilitates in higher organisms, as outlined by Bernard J. Baars. Richard Dawkins suggested that humans evolved consciousness in order to make themselves the subjects of thought. Daniel Povinelli suggests that large, tree-climbing apes evolved consciousness to take into account one's own mass when moving safely among tree branches. Consistent with this hypothesis, Gordon Gallup found that chimps and orangutans, but not little monkeys or terrestrial gorillas, demonstrated self-awareness in mirror tests.

The concept of consciousness can refer to voluntary action, awareness, or wakefulness. However, even voluntary behaviour involves unconscious mechanisms. Many cognitive processes take place in the cognitive unconscious, unavailable to conscious awareness. Some behaviours are conscious when learned but then become unconscious, seemingly automatic. Learning, especially implicitly learning a skill, can take place outside of consciousness. For example, plenty of people know how to turn right when they ride a bike, but very few can accurately explain how they actually do so.

Neural Darwinism

Neural Darwinism is a large scale theory of brain function initially proposed in 1978 by the American biologist Gerald Edelman. Edelman distinguishes between what he calls primary and secondary consciousness.

- Primary consciousness: is the ability, found in humans and some animals, to integrate observed events with memory to create an awareness of the present and immediate past of the world around them. This form of consciousness is also sometimes called "sensory consciousness". Put another way, primary consciousness is the presence of various subjective sensory contents of consciousness such as sensations, perceptions, and mental images. For example, primary consciousness includes a person's experience of the blueness of the ocean, a bird's song, and the feeling of pain. Thus, primary consciousness refers to being mentally aware of things in the world in the present without any sense of past and future; it is composed of mental images bound to a time around the measurable present.
- Secondary consciousness: is an individual's accessibility to their history and plans. The concept is also loosely and commonly associated with having awareness of one's own consciousness. The ability allows its possessors to go beyond the limits of the remembered present of primary consciousness.

Primary consciousness can be defined as simple awareness that includes perception and emotion. As such, it is ascribed to most animals. By contrast, secondary consciousness depends on and includes such features as self-reflective awareness, abstract thinking, volition and metacognition.

Edelman's theory focuses on two nervous system organizations: the brainstem and limbic systems on one side and the thalamus and cerebral cortex on the other side. The brain stem and limbic system take care of essential body functioning and survival, while the thalamocortical system receives signals from sensory receptors and sends out signals to voluntary muscles such as those of the arms and legs. The theory asserts that the connection of these two systems during evolution helped animals learn adaptive behaviors.

Other scientists have argued against Edelman's theory, instead suggesting that primary consciousness might have emerged with the basic vegetative systems of the brain. That is, the evolutionary origin might have come from sensations and primal emotions arising from sensors and receptors, both internal and surface, signaling that the well-being of the creature was immediately threatened—for example, hunger for air, thirst, hunger, pain, and extreme temperature change. This is based on neurological data showing the thalamic, hippocampal, orbitofrontal, insula, and midbrain sites are the key to consciousness of thirst. These scientists also point out that the cortex might not be as important to primary consciousness as some neuroscientists have believed. Evidence of this lies in the fact that studies show that systematically disabling parts of the cortex in animals does not remove consciousness. Another study found that children born without a cortex are conscious. Instead of cortical mechanisms, these scientists emphasize brainstem mechanisms as essential to consciousness. Still, these scientists concede that higher order consciousness does involve the cortex and complex communication between different areas of the brain.

While animals with primary consciousness have long-term memory, they lack explicit narrative, and, at best, can only deal with the immediate scene in the remembered present. While they still have an advantage over animals lacking such ability, evolution has brought forth a growing complexity in consciousness, particularly in mammals. Animals with this complexity are said to have secondary consciousness. Secondary consciousness is seen in animals with semantic capabilities, such as the four great apes. It is present in its richest form in the human species, which is unique in possessing complex language made up of syntax and semantics. In considering how the neural mechanisms underlying primary consciousness arose and were maintained during evolution, it is proposed that at some time around the divergence of reptiles into mammals and then into birds, the embryological development of large numbers of new reciprocal connections allowed rich re-entrant activity to take place between the more posterior brain systems carrying out perceptual categorization and the more frontally located systems responsible for value-category memory. The ability of an animal to relate a present complex scene to its own previous history of learning conferred an adaptive evolutionary advantage. At much later evolutionary epochs, further re-entrant circuits appeared that linked semantic and linguistic performance to categorical and conceptual memory systems. This development enabled the emergence of secondary consciousness.

Ursula Voss of the Universität Bonn believes that the theory of protoconsciousness may serve as adequate explanation for self-recognition found in birds, as they would develop secondary consciousness during REM sleep. She added that many types of birds have very sophisticated language systems. Don Kuiken of the University of Alberta finds such research interesting as well as if we continue to study consciousness with animal models (with differing types of consciousness), we would be able to separate the different forms of reflectiveness found in today's world.

For the advocates of the idea of a secondary consciousness, self-recognition serves as a critical component and a key defining measure. What is most interesting then, is the evolutionary appeal that arises with the concept of self-recognition. In non-human species and in children, the mirror test has been used as an indicator of self-awareness.

Cambridge Declaration on Consciousness

In 2012, a group of neuroscientists attending a conference on "Consciousness in Human and non-Human Animals" at the University of Cambridge in the UK, signed *The Cambridge Declaration on Consciousness.*

"The absence of a neocortex does not appear to preclude an organism from experiencing affective states. Convergent evidence indicates that non-human animals have the neuroanatomical, neurochemical, and neurophysiological substrates of conscious states along with the capacity to exhibit intentional behaviors. Consequently, the weight of evidence indicates that humans are not unique in possessing the neurological substrates that generate consciousness. Non-human animals, including all mammals and birds, and many other creatures, including octopuses, also possess these neurological substrates."

In the accompanying text they "unequivocally" asserted:

- "The field of Consciousness research is rapidly evolving. Abundant new techniques and strategies for human and non-human animal research have been developed. Consequently, more data is becoming readily available, and this calls for a periodic reevaluation of previously held preconceptions in this field. Studies of non-human animals have shown that homologous brain circuits correlated with conscious experience and perception can be selectively facilitated and disrupted to assess whether they are in fact necessary for those experiences. Moreover, in humans, new non-invasive techniques are readily available to survey the correlates of consciousness."
- "The neural substrates of emotions do not appear to be confined to cortical structures. In

fact, subcortical neural networks aroused during affective states in humans are also critically important for generating emotional behaviors in animals. Artificial arousal of the same brain regions generates corresponding behavior and feeling states in both humans and non-human animals. Wherever in the brain one evokes instinctual emotional behaviors in non-human animals, many of the ensuing behaviors are consistent with experienced feeling states, including those internal states that are rewarding and punishing. Deep brain stimulation of these systems in humans can also generate similar affective states. Systems associated with affect are concentrated in subcortical regions where neural homologies abound. Young human and non-human animals without neocortices retain these brain-mind functions. Furthermore, neural circuits supporting behavioral/electrophysiological states of attentiveness, sleep and decision making appear to have arisen in evolution as early as the invertebrate radiation, being evident in insects and cephalopod mollusks (e.g., octopus)."

- "Birds appear to offer, in their behavior, neurophysiology, and neuroanatomy a striking case of parallel evolution of consciousness. Evidence of near human-like levels of consciousness has been most dramatically observed in grey parrots. Mammalian and avian emotional networks and cognitive microcircuitries appear to be far more homologous than previously thought. Moreover, certain species of birds have been found to exhibit neural sleep patterns similar to those of mammals, including REM sleep and, as was demonstrated in zebra finches, neurophysiological patterns previously thought to require a mammalian neocortex. Magpies in particular have been shown to exhibit striking similarities to humans, great apes, dolphins, and elephants in studies of mirror self-recognition."
- "In humans, the effect of certain hallucinogens appears to be associated with a disruption in cortical feedforward and feedback processing. Pharmacological interventions in non-human animals with compounds known to affect conscious behavior in humans can lead to similar perturbations in behavior in non-human animals. In humans, there is evidence to suggest that awareness is correlated with cortical activity, which does not exclude possible contributions by subcortical or early cortical processing, as in visual awareness. Evidence that human and non-human animal emotional feelings arise from homologous subcortical brain networks provide compelling evidence for evolutionarily shared primal affective qualia."

Examples

A common image is the *scala naturae*, the ladder of nature on which animals of different species occupy successively higher rungs, with humans typically at the top. A more useful approach has been to recognize that different animals may have different kinds of cognitive processes, which are better understood in terms of the ways in which they are cognitively adapted to their different ecological niches, than by positing any kind of hierarchy.

Mammals

Dogs

Dogs were previously listed as non-self-aware animals. Traditionally, self-consciousness was evaluated via the mirror test. But dogs, as many other animals, are not visually oriented. In 2015, one study showed that the "sniff test of self-recognition" (STSR) provides significant evidences of self-awareness in dogs, and could play a crucial role in showing that this capacity is not a specific feature of only great apes, humans and a few other animals, but it depends on the way in which researchers try to verify it. According to the biologist Roberto Cazzolla Gatti, "the innovative approach to test the self-awareness with a smell test highlights the need to shift the paradigm of the anthropocentric idea of consciousness to a species-specific perspective". This study hasn't been replicated as of yet.

Birds

Grey Parrots

Research with captive grey parrots, especially Irene Pepperberg's work with an individual named Alex, has demonstrated they possess the ability to associate simple human words with meanings, and to intelligently apply the abstract concepts of shape, colour, number, zero-sense, etc. According to Pepperberg and other scientists, they perform many cognitive tasks at the level of dolphins, chimpanzees, and even human toddlers. Another notable African grey is N'kisi, which in 2004 was said to have a vocabulary of over 950 words which she used in creative ways. For example, when Jane Goodall visited N'kisi in his New York home, he greeted her with "Got a chimp?" because he had seen pictures of her with chimpanzees in Africa.

In 2011, research led by Dalila Bovet of Paris West University Nanterre La Défense, demonstrated grey parrots were able to coordinate and collaborate with each other to an extent. They were able to solve problems such as two birds having to pull strings at the same time to obtain food. In another example, one bird stood on a perch to release a food-laden tray, while the other pulled the tray out from the test apparatus. Both would then feed. The birds were observed waiting for their partners to perform the necessary actions so their behaviour could be synchronized. The parrots appeared to express individual preferences as to which of the other test birds they would work with.

Corvids

The Eurasian magpie passes the mirror test

It was recently thought that self-recognition was restricted to mammals with large brains and highly evolved social cognition, but absent from animals without a neocortex. However, in 2008, an investigation of self-recognition in corvids was conducted revealing the ability of self-recognition in the magpie. Mammals and birds inherited the same brain components from their last common ancestor nearly 300 million years ago, and have since independently evolved and formed significantly different brain types. The results of the mirror test showed that although magpies do not have a neocortex, they are capable of understanding that a mirror image belongs to their own body. The findings show that magpies respond in the mirror test in a manner similar to apes, dolphins, killer whales, pigs and elephants. This is a remarkable capability that, although not fully concrete in its determination of self-recognition, is at least a prerequisite of self-recognition. This is not only of interest regarding the convergent evolution of social intelligence, it is also valuable for an understanding of the general principles that govern cognitive evolution and their underlying neural mechanisms. The magpies were chosen to study based on their empathy/lifestyle, a possible precursor for their ability of self-awareness.

Invertebrates

Octopus travelling with shells collected for protection

Octopuses are highly intelligent, possibly more so than any other order of invertebrates. The level of their intelligence and learning capability are debated, but maze and problem-solving studies show they have both short- and long-term memory. Octopus have a highly complex nervous system, only part of which is localized in their brain. Two-thirds of an octopus' neurons are found in the nerve cords of their arms. Octopus arms show a variety of complex reflex actions that persist even when they have no input from the brain. Unlike vertebrates, the complex motor skills of octopuses are not organized in their brain using an internal somatotopic map of their body, instead using a non-somatotopic system unique to large-brained invertebrates. Some octopuses, such as the mimic octopus, move their arms in ways that emulate the shape and movements of other sea creatures.

In laboratory studies, octopuses can easily be trained to distinguish between different shapes and patterns. They reportedly use observational learning, although the validity of these findings is contested. Octopuses have also been observed to play: repeatedly releasing bottles or toys into a circular current in their aquariums and then catching them. Octopuses often escape from their aquarium and sometimes enter others. They have boarded fishing boats and opened holds to eat crabs. At least four specimens of the veined octopus (*Amphioctopus marginatus*) have been witnessed retrieving discarded coconut shells, manipulating them, and then reassembling them to use as shelter.

Animal Communication

Animal communication, process by which one animal provides information that other animals can incorporate into their decision making. The vehicle for the provision of this information is called a signal. The signal may be a sound, colour pattern, posture, movement, electrical discharge, touch, release of an odorant, or some combination of these mediums.

Animals communicate by sending and receiving signals. For example, a mother dingo (Canis lupus dingo) can communicate certain types of information to her pups by using tactile signals conveyed through grooming

Animals face daily decisions about how to behave. Choices can be as simple as a sea anemone deciding when to expand its tentacles or as complex as a male lion deciding whether to approach a reluctant mate. The decision, which may be reflexive or conscious, is guided by evolutionary biases based on alternative outcomes of choice, recent experience about likely conditions, and sensory information. An animal with access to complete information can always choose correctly. However, life is rarely so accommodating, and inputs often fail to provide complete information. Thus, communication is an important source of additional information that is incorporated into the decision-making process.

Cactus wrens (Campylorhynchus brunneicapillus) sing to communicate with other members of their species

Signals are actions or anatomical structures whose primary function is the provision of information to another animal. However, not all actions by one animal that provide information to another animal qualify as signals. The noise created by a foraging mouse and used by an owl to locate and kill the mouse is a case in point. Mice have to feed, and the noises they create while feeding (e.g., through movement and chewing) are an inadvertent result of that activity. Thus, these sounds are not a signal. In contrast, the song of a wren is not inadvertent—wrens sing solely to communicate with other birds.

Senders and Receivers

An animal that provides a signal is called a sender. The animal to which the signal is directed is the receiver. The receiver uses the signal information to help make a decision. For example, if a receiver must choose either to fight with or to flee from an opponent, it brings to this decision biases and thresholds passed on to it by successful prior generations. This information helps the receiver avoid harm and find food, shelter, and mates. Prior experience in the receiver's own life may also play a role in shaping its evaluation of the situation. If it has routinely lost fights to larger animals, a useful strategy would be to assess the size of the opponent. This may be done by using vision or other means. For example, in some cases an opponent broadcasts a low-frequency sound signal at the receiver. Because only large animals can produce low-frequency sounds, this signal provides evidence that the opponent is large. The receiver integrates its perception of the sound frequency with its prior experience and inherited avoidance of harmful situations and thus decides to flee.

In this example, the receiver can interpret the signal only if it understands that low-frequency sounds tend to be associated with large body sizes. The association between alternative signals (e.g., sounds of different frequencies) and different alternative circumstances (e.g., relative sizes of opponents) is called a code. Codes can be characterized as probabilities that a sender will emit a given signal in any given circumstance. In a perfect code, only one signal will be used in a given context, and only one context will evoke that signal. Real codes do not need to be perfect, but they do need to be good enough that a receiver attending to signals makes better decisions than if it ignored the signals and relied only on other sources of information.

Animals differ widely in the mechanisms by which they acquire signal codes. Some codes are inherited genetically. For example, the sound-producing structures of many male insects generate a limited range of sound frequencies, and the ears of females are pretuned to be most sensitive to those frequencies. In other species, senders' sounds or body odours are determined by random genetic processes, and receivers must learn which signals go with which individuals. Many songbirds have genetic limits on the range of sounds they can sing, but they can learn one or more local variants within those limits during a short period in their youth. In certain species, such as parrots or humans, both sender and receiver must learn the appropriate vocal coding, and they can continue to learn alternative coding systems throughout life.

Different contexts require different kinds of information and thus different signals. The number of signals in a species' repertoire can range from 5 or 6 in the simplest nonsocial animals to 10–20 in social insects, such as bees and ants, or to 30–40 in social vertebrates, such as wolves and primates. Most animals produce signals to attract mates and then produce additional signals to synchronize mating. Signals for mediating conflicts, including signals of aggressive intention and signals of submission, are also widespread. In addition, territorial species require signals for declaring territory ownership, and in situations in which adults guard or feed their young, both parents and offspring require signals to coordinate parental care. Social animals may use signals to coordinate group movements, to assemble dispersed group members, or to display social affiliations. Some animals have special signals that they use to share food finds, to alert others about predator attacks, and even to alert approaching predators that they have been detected. In addition, bats, oilbirds, porpoises, and electric fish use the differences between their own emitted and subsequently received signals to extract information about the ambient environment. In many of these contexts, the relevant animal signals are designed to provide a receiver with ancillary information about the identity, sex, social affiliation, and location of the sender.

Wolves are social animals, and they thus require a large repertoire of signals to communicate different kinds of information

Modes

A lamb investigates a rabbit, an example of interspecific communication using body posture and olfaction

Visual

- Gestures: The best known form of communication involves the display of distinctive body parts, or distinctive bodily movements; often these occur in combination, so a movement acts to reveal or emphasize a body part. A notable example is the presentation of a parent herring gull's bill to its chick as a signal for feeding. Like many gulls, the herring gull has a brightly coloured bill, yellow with a red spot on the lower mandible near the tip. When the parent returns to the nest with food, it stands over its chick and taps the bill on the ground; this elicits a begging response from a hungry chick (pecking at the red spot), which stimulates the parent to regurgitate food. The complete signal therefore involves a distinctive morphological feature (body part), the red-spotted bill, and a distinctive movement (tapping towards the ground) which makes the red spot highly visible to the chick. While all primates use some form of gesture, Frans de Waal concluded that apes and humans are unique in that only they use intentional gestures to communicate. He tested the hypothesis that gestures evolve into language by studying the gestures of bonobos and chimps.

A dog's visual tooth-baring threat display, accompanied by an auditory signal, growling

- Facial expression: Facial gestures play an important role in animal communication. Often a facial gesture is a signal of emotion. Dogs, for example, express anger through snarling and showing their teeth. In alarm their ears perk up, in fear the ears flatten while the dogs expose their teeth slightly and squint their eyes. Jeffrey Mogil studied the facial expressions of mice during increments of increasing pain; there were five recognizable facial expressions; orbital tightening, nose and cheek bulge, and changes in ear and whisker carriage.
- Gaze following: Social animals coordinate their communication by monitoring of each other's head and eye orientation. Such behaviour has long been recognized as an important component of communication during human development, and gaze-following has recently received much attention in animals. Studies have been conducted on apes, monkeys, dogs, birds, and tortoises, and have focused on two different tasks: "follow[ing] another's gaze into distant space" and "follow[ing] another's gaze geometrically around a visual barrier e.g. by repositioning themselves to follow a gaze cue when faced with a barrier blocking their view". The first ability has been found among a broad range of animals, while the second has been demonstrated only for apes, dogs, wolves, and corvids (ravens); attempts to demonstrate this "geometric gaze following" in marmoset and ibis gave negative results. Researchers do not yet have a clear picture of the cognitive basis of gaze following, but developmental evidence indicates that "simple" gaze following and "geometric" gaze following probably rely on different cognitive mechanisms.
- Color change: Color change can be separated into changes that occur during growth and development, and those triggered by mood, social context, or abiotic factors such as temperature. The latter are seen in many taxa. Some cephalopods, such as the octopus and the cuttlefish, have specialized skin cells (chromatophores) that can change the apparent colour, opacity, and reflectiveness of their skin. In addition to their use for camouflage, rapid changes in skin color are used while hunting and in courtship rituals. Cuttlefish may display two entirely different signals simultaneously from opposite sides of their body.

When a male cuttlefish courts a female in the presence of other males, he displays a male pattern facing the female and a female pattern facing away, to deceive other males. Some color signals occur in cycles. For example, when a female olive baboon begins to ovulate, her anogenital area swells and turns a bright red/pink. This signals to males that she is ready to mate.

- Bioluminescent communication: Communication by the production of light occurs commonly in vertebrates and invertebrates in the oceans, particularly at depths (e.g. angler fish). Two well known forms of land bioluminescence occur in fireflies and glow worms. Other insects, insect larvae, annelids, arachnids and even species of fungi possess bioluminescent abilities. Some bioluminescent animals produce the light themselves whereas others have a symbiotic relationship with bioluminescent bacteria.

Auditory

Bird calls can serve as alarms or keep members of a flock in contact, while the longer and more complex bird songs are associated with courtship and mating

Many animals communicate through vocalization. Vocal communication serves many purposes, including mating rituals, warning calls, conveying location of food sources, and social learning. In a number of species, males perform calls during mating rituals as a form of competition against other males and to signal females. Examples include hammer-headed bats, red deer, humpback whales, elephant seals, and songbirds. Other instances of vocal communication include the alarm calls of the Campbell monkey, the territorial calls of gibbons, and the use of frequency in greater spear-nosed bats to distinguish between groups. The vervet monkey gives a distinct alarm call for each of its four different predators, and the reactions of other monkeys vary appropriately according to the call. For example, if an alarm call signals a python, the monkeys climb into the trees, whereas the “eagle” alarm causes monkeys to seek a hiding place on the ground. Prairie dogs also use complex calls that signal predator differences. According to Con Slobodchikoff and others, prairie dog calls communicate the type, size, and speed of an approaching predator. Whale vocalizations have been found to have different dialects based on region.

Not all animals use vocalization as a means of auditory communication. Many arthropods rub specialized body parts together to produce sound. This is known as stridulation. Crickets and grasshoppers are well known for this, but many others use stridulation as well, including crustaceans, spiders, scorpions, wasps, ants, beetles, butterflies, moths, millipedes, and centipedes. Another means of auditory communication is the vibration of swim bladders in bony fish. The structure of swim bladders and the attached sonic muscles varies greatly across bony fish families, resulting in a wide variety of sounds. Striking body parts together can also produce auditory signals. A well-known example of this is the tail tip vibration of rattlesnakes as a warning signal. Other examples include bill clacking in birds, wing clapping in manakin courtship displays, and chest beating in gorillas.

Olfactory

Despite being the oldest method of communication, chemical communication is one of the least understood forms due in part to the sheer abundance of chemicals in our environment and the difficulty of detecting and measuring all the chemicals in a sample. The ability to detect chemicals in the environment serves many functions, a crucial one being the detection of food, a function that first arose in single-celled organisms (bacteria) living in the oceans during the early days of life on Earth. As this function evolved, organisms began to differentiate between chemicals compounds emanating from resources, conspecifics (same species; i.e., mates and kin), and heterospecifics (different species; i.e., competitors and predators). For instance, a small minnow species may do well to avoid habitat with a detectable concentration of chemical cue associated with a predator species such as northern pike. Minnows with the ability to perceive the presence of predators before they are close enough to be seen and then respond with adaptive behaviour (such as hiding) are more likely to survive and reproduce. Scent marking and scent rubbing are common forms of olfactory communication in mammals.

Electric

Electrocommunication is a rare form of communication in animals. It is seen primarily in aquatic animals, though some land mammals, notably the platypus and echidnas, sense electric fields that might be used for communication.

Weakly electric fishes provide an example of electrocommunication, together with electrolocation. These fish use an electric organ to generate an electric field, which is detected by electroreceptors. Differences in the waveform and frequency of changes in the field convey information on species, sex, and identity. These electric signals can be generated in response to hormones, circadian rhythms, and interactions with other fish. Some predators, such as sharks and rays, are able to eavesdrop on these electrogenic fish through passive electroreception.

Touch

Touch is a key factor in many social interactions. Here are some examples:

- Fighting: In a fight, touch may be used to challenge an opponent and to coordinate movements during the fight. It may also be used by the loser to indicate submission.
- Mating: Mammals often initiate mating by grooming, stroking or rubbing against each other. This provides the opportunity to apply chemical signals and to assess those excreted by the potential mate. Touch may also announce the intention of the male to mount the female, as when a male kangaroo grabs the tail of a female. During mating, touch stimuli are important for pair positioning, coordination and genital stimulation.

- Social integration: Touch is widely used for social integration, a use that is typified by the social grooming of one animal by another. Social grooming has several functions; it removes parasites and debris from the groomed animal, it reaffirms the social bond or hierarchical relationship between the animals, and it gives the groomer an opportunity to examine olfactory cues on the groomed individual, perhaps adding additional ones. This behaviour has been observed in social insects, birds and mammals.
- Foraging: Some ant species recruit fellow workers to new food finds by first tapping them with their antennae and forelegs, then leading them to the food source while keeping physical contact. Another example of this is the waggle dance of honey bees.
- Huddling: Prolonged physical contact or huddling also serves social integration. Huddling promotes heat exchange, together with the transfer of olfactory or tactile information. Some organisms live in constant contact in a colony, for example colonial corals. When individuals are linked tightly in this way an entire colony can react on the aversive or alarm movements made by only a few individuals. In several herbivorous insect nymphs and larvae, aggregations where there is prolonged contact play a major role in group coordination. These aggregations may take the form of a procession or a rosette.

Seismic

Seismic communication is the exchange of information using self-generated vibrational signals transmitted via a substrate such as the soil, water, spider webs, plant stems, or a blade of grass. This form of communication has several advantages, for example it can be sent regardless of light and noise levels, and it usually has a short range and short persistence, which may reduce the danger of detection by predators. The use of seismic communication is found in many taxa, including frogs, kangaroo rats, mole rats, bees, nematode worms, and others. Tetrapods usually make seismic waves by drumming on the ground with a body part, a signal that is sensed by the sacculus of the receiver. The sacculus is an organ in the inner ear containing a membranous sac that is used for balance, but can also detect seismic waves in animals that use this form of communication. Vibrations may be combined with other sorts of communication.

Thermal

A number of different snakes have the ability to sense infrared (IR) thermal radiation, which allows these reptiles to derive thermal images from the radiant heat emitted by predators or prey at wavelengths between 5 and 30 μm. The accuracy of this sense is such that a blind rattlesnake can target its strike to the vulnerable body parts of a prey animal. It was previously thought that the pit organs evolved primarily as prey detectors, but it is now believed that they may also be used to control body temperature.

The facial pits enabling thermoregulation underwent parallel evolution in pitvipers and some boas and pythons, having evolved once in pitvipers and multiple times in boas and pythons. The electrophysiology of the structure is similar between lineages, but it differs in gross structure anatomy. Most superficially, pitvipers possess one large pit organ on either side of the head, between the eye and the nostril (loreal pit), while boas and pythons have three or more comparatively smaller pits lining the upper and sometimes the lower lip, in or between the scales. Those of the pitvipers are the more advanced, having a suspended sensory membrane as opposed to a simple pit structure. Within the family Viperidae, the pit organ is seen only in the subfamily Crotalinae: the pitvipers. Despite the detection of IR radiation, the pits' IR mechanism is dissimilar to photoreceptors; while photoreceptors detect light via photochemical reactions, the protein in the facial pits of snakes is a temperature sensitive ion channel. It senses infrared signals through a mechanism involving warming of the pit organ, rather than chemical reaction to light. This is consistent with the thin pit membrane, which allows incoming IR radiation to quickly and precisely warm a given ion channel and trigger a nerve impulse, as well as vascularize the pit membrane to rapidly cool the ion channel back to its original "resting" or "inactive" temperature.

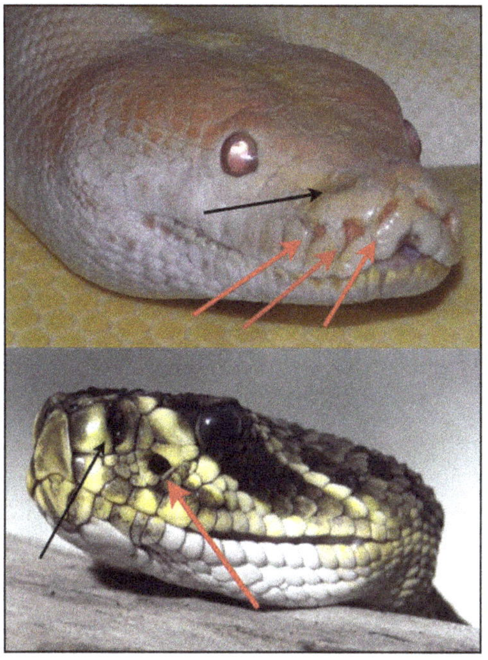

A python (top) and rattlesnake illustrating the positions of the pit organs. Red arrows indicate the pit organs whereas black arrows indicate the nostril

Common vampire bats (*Desmodus rotundus*) have specialized IR sensors in their nose-leaf. Vampire bats are the only mammals that feed exclusively on blood. The IR sense enables Desmodus to localize homeothermic animals such as cattle and horses within a range of about 10 to 15 cm. This infrared perception may be used in detecting regions of maximal blood flow on targeted prey.

Bee Learning and Communication

Honey bees are sensitive to odors (including pheromones), tastes, and colors, including ultraviolet. They can demonstrate capabilities such as color discrimination through classical and operant conditioning and retain this information for several days at least; they communicate the location and nature of sources of food; they adjust their foraging to the times at which food is available; they may even form cognitive maps of their surroundings.

Bees learn and communicate in a variety of ways

Learning

Learning is essential for efficient foraging. Honey bees are unlikely to make many repeat visits if a plant provides little in the way of reward. A single forager will visit different flowers in the morning and, if there is sufficient attraction and reward in a particular kind of flower, she will make visits to that type of flower for most of the day, unless the plants stop producing reward or weather conditions change. Honey bees are quite adept at associative learning, and many of the standard phenomena of classical conditioning take the same form in honey bees as they do in the vertebrates that are the more usual subjects of such experiments.

Swarming bees require good communication to all congregate in the same place

Foragers were trained to enter a simple Y-shaped maze that had been marked at the entrance with a particular color. Inside the maze was a branching point where the bee was required to choose between two paths. One path, which led to the food reward, was marked with the same color that had been used at the entrance to the maze, while the other was marked with a different color. Foragers learned to choose the correct path, and continued to do so when a different kind of marker (black and white stripes oriented in various directions) was substituted for the colored markers. When the experimental conditions were reversed, rewarding bees for choosing the inner passage marked with a symbol that was different from the entrance symbol, the bees again learned to choose the correct path. Extending the length of the tunnel to increase the time between seeing the one marker indicating the correct path and a second marker identifying the correct path show that the bees can retain the information in their visual working memory for about 5 seconds, equivalent to the short-term memory of birds.

Color Learning in Honeybees

One of the most common ways that honey bees, *Apis mellifera*, demonstrate associative learning is in the context of color recognition and discrimination tasks. Just as vertebrate species such as mice or pigeons that can be trained to perform associative learning tasks, honey bees make excellent subjects for tasks involving discrimination and color memory. Beginning in the early 1900s, scientists Karl von Frisch and later Randolf Menzel began asking questions about the existence, learning rates, memory, and timing of color vision in bees.

Color Discrimination

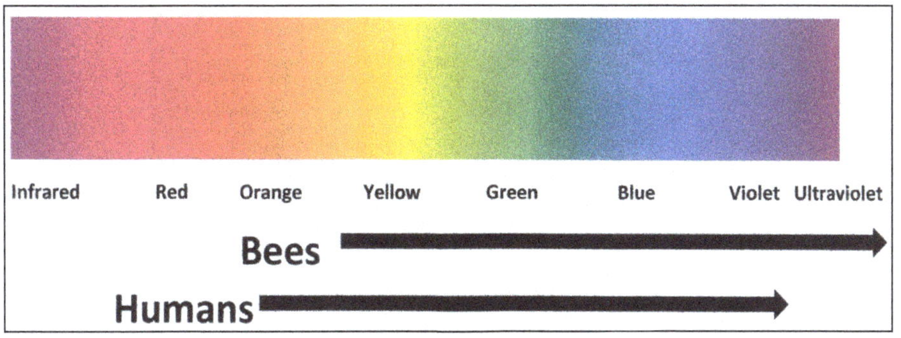

Bee Vision vs. Human Vision

The Austrian zoologist Karl von Frisch began the exploration of color vision in honey bees when he asked the first question in 1919: does color vision in bees exist? By making use of bees associative learning abilities he performed an elegant experiment to show that honey bees were in fact capable of color discrimination.

To test color vision, von Frisch first trained his honey bees to feed from a small dish filled with a nectar-like sugar water. This dish was placed over a small piece of blue colored cardboard so that the color was visible to the bees as they fed. Once the bees had become accustomed to the blue cardboard, von Frisch surrounded the blue piece of cardboard with other identically sized pieces in varying shades of gray and placed small dishes over each piece. If bees could not discriminate between colors, they would be unable to distinguish the blue piece from the many gray toned pieces. In the case that bees did not have color vision then, von Frisch predicted that the bees would visit the gray and blue pieces with equal frequency, as they would not be able to tell a difference between them.

When he allowed bees access to the dishes, however, he found that the vast majority of the bees flew directly to the blue piece of cardboard on which they had previously obtained their sugar-water reward. The bees largely ignored the gray pieces which had not been rewarded. This directed exploration and targeting of the blue cardboard showed the bees could indeed discriminate between the gray and blue shades, showing that bees do possess color vision. Von Frisch repeated this same basic experiment to show that bees produced the same results with other colors like violet and yellow. Later other researchers were able to apply this excellent experimental design to other vertebrates as well, making it an invaluable insight into testing color vision in many organisms.

Color Learning Rates and Preferences

After von Frisch's initial studies, the German scientist Randolf Menzel continued the study of color vision in honey bees and performed more detailed tests. He was curious about which colors honey bees would be able to learn fastest and whether or not bees had a greater aptitude for learning certain colors.

He used lights of varying color and intensity to illuminate circles of light on a solid surface. This set up was similar to the pieces of colored cardboard employed by von Frisch, but by using light instead of cardboard, Menzel was able to change the intensity and color of light easily. He could simply adjust the projection of the light to create a wide variety of different experimental set-ups.

Honey bee collecting pollen

To test the intricacies of the bee color vision von Frisch had established, Menzel performed an experiment that aimed to test bees ability to distinguish between two different colors. To do this, Menzel used a projected circle of colored light surrounding a small dish that could hold a sugar-water reward. Menzel then projected a second circle of differently colored light surrounding a second dish some distance away from the first. Next, a single bee was placed equidistant between these two different lights and allowed to choose which dish to search for a sugar-water treat. Only one of the colored light circles surrounded a dish that contained sugar-water; the other was empty. Menzel was then able to measure how quickly the bees learned to preferentially search only the rewarded light and to ignore the dish surrounded by unrewarded light.

The results of the experiment showed that bees did not learn to discriminate between all color pairs equally well. The fastest rate of learning was when violet light was rewarded. The color that the bees had the most difficulty learning was green, and all other colors fell somewhere in between. This evidence of inherent bias is evolutionarily reasonable given that color vision in bees allows them to distinguish between different nectar-bearing flowers, much like the rewarded dishes. As more flowers are purple than green it makes sense that bees would be more sensitive to colors likely to result in nourishment.

Color Memory

After this work on color preferences, Menzel extended his experiments to study memory in honey bee color vision. He wanted to know how many trials were necessary for honey bees to reliably choose a previously rewarded color when presented with several choices for potential rewards and how long honey bees could retain information about which color would be rewarded.

To test these questions, Menzel performed a variety of experiments. First, he presented individual bees with a sugar reward on a colored background for just a single trial. He then kept these bees in small cages for several days without any further trials. After a few days, he presented each bee with several dishes each on a different colored background at once. One of the colors was the same as that used during the initial trial. The others were novel, unrewarded colors. Amazingly, after just one trial and several days without any exposure to the rewarded color, bees correctly chose to explore the color used in the first trial more than fifty-percent of the time.

Menzel then repeated this experiment with another group of bees, keeping all factors the same except that in the second round of testing he gave the bees three initial trials with the rewarded color instead of just one. After several days in confinement when the bees were presented with a choice of colors just as in the first experiment, they virtually always chose the color that had been used during the first three trials.

This ability to retain information about color-linked rewards over a period of several days and after only minimal exposure to the colored background indicates the great strength of honeybees memory with respect to color vision.

Timing in Color Learning

One of von Frisch's students, Elizabeth Opfinger, observed that bees would learn color when approaching a feeder. Menzel took this question further: when do bees register and learn color? He wanted to know if bees registered color before, during, or after receiving their sugar-water reward. In order to explore this intriguing question, Menzel displayed the color beneath a rewarded dish at different stages of the honey bee feeding process: during approach, feeding and departure.

The outcome of this experiment revealed that bees register color during both the approach and feeding stages of the exposure process. In order for a bee to accurately remember a given color, it must be present for approximately five seconds in total. Although it varies slightly, Menzel and his colleagues found that bees usually remember best when the stimulus is present for about three seconds during the approach and two seconds after landing and beginning to feed.

Neurobiology of Color Vision

Color vision in honey bees can also be examined from a neurobiological perspective in terms of the structure and organization of their compound eyes.

In 1975 Menzel published a seminal paper describing the morphology and spectral sensitivity of the honey bee eye. He examined color-coding the honey bee retina by using a technique to mark individual cells with a fluorescent dye and record from these cells as single units. Such fine structure analysis allowed him to determine that there are three types of receptors in the honey bee eye.

1) UV receptors, 2) blue receptors, and 3) green receptors. The three receptors are dominated by three rhodopsin-like pigments. These pigments have maximal absorbance at wavelengths corresponding to 350 nm, 440 nm, and 540 nm.

Western honey bee

As the cells were examined in detail, certain features were distinguishable for each type of receptor cell. UV cells were found to form the longest visual fibres. These long visual fibers penetrated the lamina with arborizations, a tree-like branching of the fibers and spines. Blue and green receptor cells have more shallow fibers.

Interestingly, Menzel found that most of the cells he studied had secondary sensitivities that corresponded to wavelength regions at which the other two receptor types were maximally active. He used spectral efficiency experiments to show that such corresponding wavelength receptivity is the result of electric coupling.

Communication

Foragers communicate their floral findings in order to recruit other worker bees of the hive to forage in the same area. The factors that determine recruiting success are not completely known but probably include evaluations of the quality of nectar and/or pollen brought in.

There are two main hypotheses to explain how foragers recruit other workers—the "waggle dance" or "dance language" theory and the "odor plume" theory. The dance language theory is far more widely accepted, and has far more empirical support. The theories also differ in that the former allows for an important role of odor in recruitment (i.e., effective recruitment relies on dance *plus* odor), while the latter claims that the dance is essentially irrelevant (recruitment relies on odor alone). The academic debate between these two theories is extremely polarized and often hostile.

Dance Communication

It has long been known that successfully foraging Western honey bees perform a dance on their return to the hive, known as *waggle dance*, indicating that food is farther away, while the *round dance* is a short version of the *waggle dance*, indicating that food is nearby. The laden forager dances on the comb in a circular pattern, occasionally crossing the circle in a zig-zag or waggle pattern. Aristotle described this behaviour in his *Historia Animalium*. It was thought to attract the attention of other bees.

Eight-Shaped waggle dance of the honeybee (Apis mellifera). A waggle run oriented 45° to the right of 'up' on the vertical comb indicates a food source 45° to the right of the direction of the sun outside the hive. The abdomen of the dancer appears blurred because of the rapid motion from side to side

In 1947, Karl von Frisch correlated the runs and turns of the dance to the distance and direction of the food source from the hive. The orientation of the dance correlates to the relative position of the sun to the food source, and the length of the waggle portion of the run is correlated to the distance from the hive. Also, the more vigorous the display is, the better the food. There is no evidence that this form of communication depends on individual learning. Honeybees detect the dances of conspecifics by sensing near field sound and electric fields using the Johnston's organ.

Von Frisch performed a series of experiments to validate his theory. He was awarded the Nobel Prize in Physiology or Medicine in 1973 for his discoveries.

One of the most important lines of evidence on the origin and utility of the dance is that all of the known species and races of honey bees exhibit the behavior, but details of its execution vary among the different species. For example, in *Apis florea* and *Apis andreniformis* (the "dwarf honeybees") the dance is performed on the dorsal, horizontal portion of the nest, which is exposed. The runs and dances point directly toward the resource in these species. Each honey bee species has a characteristically different correlation of "waggling" to distance, as well. Such species-specific behavior suggests that this form of communication does not depend on learning but is rather determined genetically. It also suggests how the dance may have evolved.

Various experiments document that changes in the conditions under which the dance is performed lead to characteristic changes in recruitment to external resources, in a manner consistent with von Frisch's original conclusions. Experiments with robotic dummies were indeed to induce some recruitment, which should not have been possible if the dance contains no information. Various experimental results demonstrate that the dance does convey information, but the use of this information may be context-dependent, and this may explain why the results of earlier studies were inconsistent. Researchers have also discovered other forms of honeybee dance communication, such as the tremble dance.

Odor Plume

While many researchers believe that bee dances give enough information to locate resources, proponents of the odor plume theory argue that the dance gives little, or no actual guidance to a nectar source. They argue that bees instead are primarily recruited by odor. The purpose of the dance is simply to gain attention to the returning worker bee so she can share the odor of the nectar with other workers who will then follow the odor trail to the source. Most scientists agree that odor is used in recruitment to resources, but they differ strongly in opinion as to the information content of the dance.

The primary lines of evidence used by the odor plume advocates are:

1. experiments with odorless sugar sources which show that worker bees are unable to recruit to those sources and,
2. logical difficulties of a small-scale dance (a few centimeters across) giving directions precise enough to hold the other bees on course during a flight that could be several kilometers long. Misreading by even a few degrees would lead the bee off course by hundreds of meters at the far end.

Neither of these points invalidate the dance theory, but simply suggest that odor might be involved, which is indeed conceded by all proponents of dance theory. Critics of the odor plume theory counter that most natural nectar sources are relatively large—orchards or entire fields— so, precision may not be necessary or even desirable. They have also challenged the reproducibility of the odorless source experiment.

Odor learning in bees is usually tested by the proboscis extension reflex. Significant to the argument are the experiments of William F. Towne, of the Kutztown University in Pennsylvania, in which hives are moved to "mirror image" terrain settings, and the bees are thereby fooled into dancing about the wrong location for a nectar source. Foragers were successfully recruited to the wrong location, but only when the sun was obscured by clouds, forcing them to rely on terrain-based navigation rather than "solar ephemeris"-based navigation. As the cloud cover broke up, more and more bees corrected their dances to indicate the actual location of nectar, and forager visits shifted to the correct location.

Odor is an essential and even necessary at various stages of the recruitment process, including once a recruited forager reaches the vicinity of the resource while some scientists think that dancing may be a simple idiothetic movement that conveys no information. Others see the dance as conveying information, but doing it poorly compared to other means and potentially used backup approach.

Note: much of the research on the two competing hypotheses of communication has been restricted to Western honey bees. Other species of *Apis* use variants on the same theme, and other types of bees use other methods altogether.

Trophallaxis

The exchange of food, trophallaxis, can be used to communicate the quality of a food source, temperature, a need for water, and the condition of the queen.

Primer Pheromones

Research that was published in November 2004, by scientists under the leadership of Zachary Huang, Michigan State University indicates that so called primer pheromones play an important part in how a honey bee colony adjusts its distribution of labor most beneficially. In order to survive as a bee colony of sometimes 50,000–100,000 individual bees, the communal structure has to be adaptable to seasonal changes and the availability of food. The division of labor has to adjust itself to the resources available from foraging. While the division of labor in a bee colony is quite complex, the work can be roughly seen as work inside the hive and outside the hive. Younger bees play a role inside the hive while older bees play a role outside the hive mostly as foragers. Huang's team found that forager bees gather and carry a chemical called ethyl oleate in the stomach. The forager bees feed this primer pheromone to the worker bees, and the chemical keeps them in a nurse bee state. The pheromone prevents the nurse bees from maturing too early to become forager bees. As forager bees die off, less of the ethyl oleate is available and nurse bees more quickly mature to become foragers. It appears that this control system is an example of decentralized decision making in the bee colony.

Other bees like *Trigona corvina* rely on pheromones for much of their communication with nest mates and rivals. They produce pheromones from their labial glands. The function of signaling depends on the profitability, but they commonly will scent mark a food source either for self-orientation, to deter rivals or to direct a nest mate to the resource. Once an individual finds a good food source, they will return to the same source for many days. If an individual detects the scent of a rival bee, they will avoid the plant in order to avoid conflict and to save time. It has also been shown that pheromones are a method of sexual selection between male drones and queens.

Cognition

Experiments by James Gould suggest that honey bees may have a cognitive map for information they have learned, and utilize it when communicating.

In one test reported in a 1983 issue of *Science News*, he moved a supply of sugar water 25% further away from a hive each day. The bees communicated to each other as usual on its location. Then he placed the sugar water on a boat anchored in the middle of a small lake. When scouts returned to the hive to communicate their find, other bees refused to go with them, even though they frequently flew over the lake to reach pollen sources on the opposite shore.

In another test related in the August 1986 issue of *Discover* ("A Honey of a Question: Are Bees Intelligent?"), Gould lured some bees to a dish of artificial nectar, then gradually moved it farther from the hive after they became accustomed to it. He marked the trained bees, placed them in a darkened jar, and relocated them to a spot where the hive was still visible, but not the dish. When released one by one, the bees would appear disoriented for a few seconds, then fly directly for the covert dish. Seventy-three of 75 bees reached it in about 28 seconds. They apparently accomplished this feat by devising a new flight path based on a cognitive map of visible landmarks.

Comparative Psychology

Comparative psychology refers to the scientific study of the behavior and mental processes of non-human animals, especially as these relate to the phylogenetic history, adaptive significance, and development of behavior. Research in this area addresses many different issues, uses many different methods and explores the behavior of many different species from insects to primates.

Comparative psychology is sometimes assumed to emphasize cross-species comparisons, including those between humans and animals. However, some researchers feel that direct comparisons should not be the sole focus of comparative psychology and that intense focus on a single organism to understand its behavior is just as desirable; if not more so. Donald Dewsbury reviewed the works of several psychologists and their definitions and concluded that the object of comparative psychology is to establish principles of generality focusing on both proximate and ultimate causation.

Using a comparative approach to behavior allows one to evaluate the target behavior from four different, complementary perspectives, developed by Niko Tinbergen. First, one may ask how pervasive the behavior is across species (i.e. how common is the behavior between animal species?). Second, one may ask how the behavior contributes to the lifetime reproductive success of the individuals demonstrating the behavior (i.e. does the behavior result in animals producing more offspring than animals not displaying the behavior)? Theories addressing the ultimate causes of behavior are based on the answers to these two questions.

Third, what mechanisms are involved in the behavior (i.e. what physiological, behavioral, and environmental components are necessary and sufficient for the generation of the behavior)? Fourth, a researcher may ask about the development of the behavior within an individual (i.e. what maturational, learning, social experiences must an individual undergo in order to demonstrate a behavior)? Theories addressing the proximate causes of behavior are based on answers to these two questions.

Species Studied

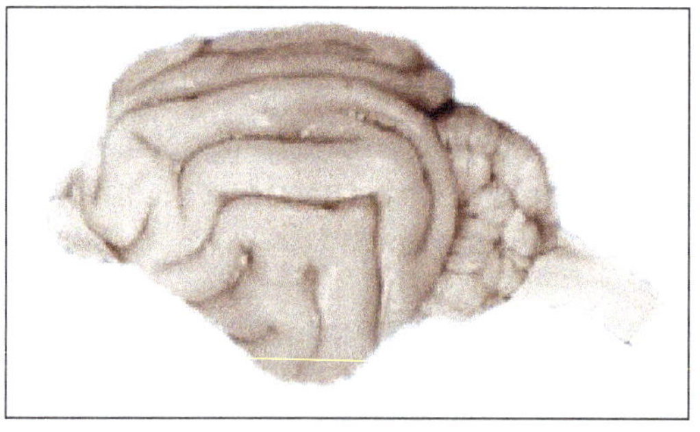

the brain of a cat

A wide variety of species have been studied by comparative psychologists. However, a small number have dominated the scene. Ivan Pavlov's early work used dogs; although they have been the subject of occasional studies, since then they have not figured prominently. Increasing interest in the study of abnormal animal behavior has led to a return to the study of most kinds of domestic animal. Thorndike began his studies with cats, but American comparative psychologists quickly shifted to the more economical rat, which remained the almost invariable subject for the first half of the 20th century and continues to be used.

Skinner introduced the use of pigeons, and they continue to be important in some fields. There has always been interest in studying various species of primate; important contributions to social and developmental psychology were made by Harry F. Harlow's studies of maternal deprivation in rhesus monkeys. Cross-fostering studies have shown similarities between human infants and infant chimpanzees. Kellogg and Kellogg (1933) aimed to look at heredity and environmental effects of young primates. They found that a cross-fostered chimpanzee named Gua was better at recognizing human smells and clothing and that the Kelloggs' infant (Donald) recognised humans better by their faces. The study ended 9 months after it had begun, after the infant began to imitate the noises of Gua.

Nonhuman primates have also been used to show the development of language in comparison with human development. For example, Gardner (1967) successfully taught the female chimpanzee Washoe 350 words in American Sign Language. Washoe subsequently passed on some of this teaching to her adopted offspring, Loulis. A criticism of Washoe's acquisition of sign language focused on the extent to which she actually understood what she was signing. Her signs may have just based on an association to get a reward, such as food or a toy. Other studies concluded that apes do not understand linguistic input, but may form an intended meaning of what is being communicated. All great apes have been reported to have the capacity of allospecific symbolic production.

Interest in primate studies has increased with the rise in studies of animal cognition. Other animals thought to be intelligent have also been increasingly studied. Examples include various species of corvid, parrots — especially the grey parrot — and dolphins. Alex (Avian Learning EXperiment) is a well known case study (1976–2007) which was developed by Pepperberg, who found that the African gray parrot Alex did not only mimic vocalisations but understood the concepts of same and different between objects. The study of non-human mammals has also included the study of dogs. Due to their domestic nature and personalities, dogs have lived closely with humans, and parallels in communication and cognitive behaviours have therefore been recognised and further researched. Joly-Mascheroni and colleagues (2008) demonstrated that dogs may be able to catch human yawns and suggested a level of empathy in dogs, a point that is strongly debated. Pilley and Reid found that a Border Collie named Chaser was able to successfully identify and retrieve 1022 distinct objects/toys.

Animal Cognition

Researchers who study animal cognition are interested in understanding the mental processes that control complex behavior, and much of their work parallels that of cognitive psychologists working with humans. For example, there is extensive research with animals on attention, categorization, concept formation, memory, spatial cognition, and time estimation. Much research in these and other areas is related directly or indirectly to behaviors important to survival in natural settings, such as navigation, tool use, and numerical competence. Thus, comparative psychology and animal cognition are heavily overlapping research categories.

Disorders of Animal Behavior

Veterinary surgeons recognize that the psychological state of a captive or domesticated animal must be taken into account if its behavior and health are to be understood and optimized.

Common causes of disordered behavior in captive or pet animals are lack of stimulation, inappropriate stimulation, or overstimulation. These conditions can lead to disorders, unpredictable and unwanted behavior, and sometimes even physical symptoms and diseases. For example, rats who are exposed to loud music for a long period will ultimately develop unwanted behaviors that have been compared with human psychosis, like biting their owners.

The way dogs behave when understimulated is widely believed to depend on the breed as well as on the individual animal's character. For example, huskies have been known to ruin gardens and houses if they are not allowed enough activity. Dogs are also prone to psychological damage if they are subjected to violence. If they are treated very badly, they may become dangerous.

The systematic study of disordered animal behavior draws on research in comparative psychology, including the early work on conditioning and instrumental learning, but also on ethological studies of natural behavior. However, at least in the case of familiar domestic animals, it also draws on the accumulated experience of those who have worked closely with the animals.

Human-animal Relationships

The relationship between humans and animals has long been of interest to anthropologists as one pathway to an understanding the evolution of human behavior. Similarities between the behavior of humans and animals have sometimes been used in an attempt to understand the evolutionary significance of particular behaviors. Differences in the treatment of animals have been said to reflect a society's understanding of human nature and the place of humans and animals in the scheme of things. Domestication has been of particular interest. For example, it has been argued that, as animals became domesticated, humans treated them as property and began to see them as inferior or fundamentally different from humans.

Ingold remarks that in all societies children have to learn to differentiate and separate themselves from others. In this process, strangers may be seen as "not people," and like animals. Ingold quoted Sigmund Freud: "Children show no trace of arrogance which urges adult civilized men to draw a hard-and-fast line between their own nature and that of all other animals. Children have no scruples over allowing animals to rank as their full equals." With maturity however, humans find it hard to accept that they themselves are animals, so they categorize, separating humans from animals, and animals into wild animals and tame animals, and tame animals into house pets and livestock. Such divisions can be seen as similar to categories of humans: who is part of a human community and someone who isn't, that is, the outsider.

The New York Times ran an article that showed the psychological benefits of animals, more specifically of children with their pets. It's been proven that having a pet does in fact improve kids' social skills. In the article, Dr. Sue Doescher, a psychologist involved in the study, stated, "It made the children more cooperative and sharing." It was also shown that these kids were more confident with themselves and able to be more empathic with other children.

Furthermore, in an edition of *Social Science and Medicine* it was stated, "A random survey of 339 residents from Perth, Western Australia were selected from three suburbs and interviewed by telephone. Pet ownership was found to be positively associated with some forms of social contact and interaction, and with perceptions of neighborhood friendliness. After adjustment for demographic variables, pet owners scored higher on social capital and civic engagement scales." Results like these let us know that owning a pet provides opportunities for neighborly interaction, among many other chances for socialization among people.

Emotion in Animals

Charles Darwin was one of the first scientists to write about the existence and nature of emotions in animals. His observational (and sometimes anecdotal) approach has developed into a more robust, hypothesis-driven, scientific approach. General hypotheses relating to correlates between humans and animals also support the claim that animals may feel emotions and that human emotions evolved from the same mechanisms. Several tests, such as cognitive bias tests and learned helplessness models, have been developed. Cognitive biases (feelings of optimism or pessimism) have been shown in a wide range of species including rats, dogs, cats, rhesus macaques, sheep, chicks, starlings, pigs and honeybees.

A drawing of a cat by T. W. Wood in Charles Darwin's book *The Expression of the Emotions in Man and Animals*, described as acting "in an affectionate frame of mind"

Some behaviourists, such as John B. Watson, claim that stimulus–response models provide a sufficient explanation for animal behaviours that have been described as emotional, and that all behaviour, no matter how complex, can be reduced to a simple stimulus-response association. Watson described that the purpose of psychology was "to predict, given the stimulus, what reaction will take place; or given the reaction, state what the situation or stimulus is that has caused the reaction".

Definitions, and Differentiation

The word "emotion" dates back to 1579, when it was adapted from the French word *émouvoir*, which means "to stir up". However, the earliest precursors of the word likely date back to the very origins of language.

Emotions have been described as discrete and consistent responses to internal or external events which have a particular significance for the organism. Emotions are brief in duration and consist of a coordinated set of responses, which may include physiological, behavioural, and neural mechanisms. Emotions have also been described as the result of evolution because they provided good solutions to ancient and recurring problems that faced ancestors.

Laterality

It has been proposed that negative, withdrawal-associated emotions are processed predominantly by the right hemisphere, whereas the left hemisphere is largely responsible for processing positive, approach-related emotions. This has been called the "laterality-valence hypothesis".

Basic and Complex Human Emotions

In humans, a distinction is sometimes made between "basic" and "complex" emotions. Six emotions have been classified as basic: anger, disgust, fear, happiness, sadness and surprise. Complex emotions would include contempt, jealousy and sympathy. However, this distinction is difficult to maintain, and animals are often said to express even the complex emotions.

Background

Behaviourist Approach

A squirrel communicating with its pup

Prior to the development of animal sciences such as comparative psychology and ethology, interpretation of animal behaviour tended to favour a minimalistic approach known as behaviourism. This approach refuses to ascribe to an animal a capability beyond the least demanding that would explain a behaviour; anything more than this is seen as unwarranted anthropomorphism. The behaviourist argument is, why should humans postulate consciousness and all its near-human implications in animals to explain some behaviour, if mere stimulus-response is a sufficient explanation to produce the same effects?

The cautious wording of Dixon exemplifies this viewpoint:

> "Recent work in the area of ethics and animals suggests that it is philosophically legitimate to ascribe emotions to animals. Furthermore, it is sometimes argued that emotionality is a morally relevant psychological state shared by humans and non humans. What is missing from the philosophical literature that makes reference to emotions in animals is an attempt to clarify and defend some particular account of the nature of emotion, and the role that emotions play in a characterization of human nature. I argue in this paper that some analyses of emotion are more credible than others. Because this is so, the thesis that humans and nonhumans share emotions may well be a more difficult case to make than has been recognized thus far."

Moussaieff Masson and McCarthy describe a similar view (with which they disagree):

> "While the study of emotion is a respectable field, those who work in it are usually academic psychologists who confine their studies to human emotions. The standard reference work, *The Oxford Companion to Animal Behaviour*, advises animal behaviourists that "One is well advised to study the behaviour, rather than attempting to get at any underlying emotion. There is considerable uncertainty and difficulty related to the interpretation and ambiguity of emotion: an animal may make certain movements and sounds, and show certain brain and chemical signals when its body is damaged in a particular way. But does this mean an animal feels—is *aware* of—pain as we are, or does it merely mean it is programmed to act a certain way with certain stimuli? Similar questions can be asked of any activity an animal (including a human) might undertake, in principle. Many scientists regard all emotion and cognition (in humans and animals) as having a purely mechanistic basis."

Because of the philosophical questions of consciousness and mind that are involved, many scientists have stayed away from examining animal and human emotion, and have instead studied measurable brain functions through neuroscience.

Comparative Approach

In 1903, C. Lloyd Morgan published Morgan's Canon, a specialised form of Occam's razor used in ethology, in which he stated:

> "In no case is an animal activity to be interpreted in terms of higher psychological processes, if it can be fairly interpreted in terms of processes which stand lower in the scale of psychological evolution and development."

Darwin's Approach

Charles Darwin initially planned to include a chapter on emotion in *The Descent of Man* but as his ideas progressed they expanded into a book, *The Expression of the Emotions in Man and Animals*. Darwin proposed that emotions are adaptive and serve a communicative and motivational function, and he stated three principles that are useful in understanding emotional expression: First, *The Principle of Serviceable Habits* takes a Lamarckian stance by suggesting that emotional expressions that are useful will be passed on to the offspring. Second, *The Principle of Antithesis* suggests that some expressions exist merely because they oppose an expression that is useful. Third, *The Principle of the Direct Action of the Excited Nervous System on the Body* suggests that emotional expression occurs when nervous energy has passed a threshold and needs to be released.

Darwin saw emotional expression as an outward communication of an inner state, and the form of that expression often carries beyond its original adaptive use. For example, Darwin remarks that humans often present their canine teeth when sneering in rage, and he suggests that this means that a human ancestor probably utilized their teeth in aggressive action. A domestic dog's simple tail wag may be used in subtly different ways to convey many meanings as illustrated in Darwin's *The Expression of the Emotions in Man and Animals* published in 1872.

Examples of tail position indicating different emotions in dogs

"Small dog watching a cat on a table"

"Dog approaching another dog with hostile intentions"

"Dog in a humble and affectionate frame of mind"

"Half-bred shepherd dog"

"Dog caressing his master"

Anecdotal Approach

Evidence for emotions in animals has been primarily anecdotal, from individuals who interact with pets or captive animals on a regular basis. However, critics of animals having emotions often suggest that anthropomorphism is a motivating factor in the interpretation of the observed behaviours. Much of the debate is caused by the difficulty in defining emotions and the cognitive requirements thought necessary for animals to experience emotions in a similar way to humans. The problem is made more problematic by the difficulties in testing for emotions in animals. What is known about human emotion is almost all related or in relation to human communication.

Scientific Approach

In recent years, the scientific community has become increasingly supportive of the idea of emotions in animals. Scientific research has provided insight into similarities of physiological changes between humans and animals when experiencing emotion.

Much support for animal emotion and its expression results from the notion that feeling emotions doesn't require significant cognitive processes, rather, they could be motivated by the processes to act in an adaptive way, as suggested by Darwin. Recent attempts in studying emotions in animals have led to new constructions in experimental and information gathering. Professor Marian Dawkins suggested that emotions could be studied on a functional or a mechanistic basis. Dawkins suggests that merely mechanistic or functional research will provide the answer on its own, but suggests that a mixture of the two would yield the most significant results.

Functional

Functional approaches rely on understanding what roles emotions play in humans and examining that role in animals. A widely used framework for viewing emotions in a functional context is that described by Oatley and Jenkins who see emotions as having three stages: (i) appraisal in which there is a conscious or unconscious evaluation of an event as relevant to a particular goal. An emotion is positive when that goal is advanced and negative when it is impeded (ii) action readiness where the emotion gives priority to one or a few kinds of action and may give urgency to one so that it can interrupt or compete with others and (iii) physiological changes, facial expression and then behavioural action. The structure, however, may be too broad and could be used to include all the animal kingdom as well as some plants.

Mechanistic

The second approach, mechanistic, requires an examination of the mechanisms that drive emotions and search for similarities in animals.

The mechanistic approach is utilized extensively by Paul, Harding and Mendl. Recognizing the difficulty in studying emotion in non-verbal animals, Paul et al. demonstrate possible ways to better examine this. Observing the mechanisms that function in human emotion expression, Paul et al. suggest that concentration on similar mechanisms in animals can provide clear insights into the animal experience. They noted that in humans, cognitive biases vary according to emotional state and suggested this as a possible starting point to examine animal emotion. They propose that researchers may be able to use controlled stimuli which have a particular meaning to trained animals to induce particular emotions in these animals and assess which types of basic emotions animals can experience.

Cognitive Bias Test

Is the glass half empty or half full?

A cognitive bias is a pattern of deviation in judgment, whereby inferences about other animals and situations may be drawn in an illogical fashion. Individuals create their own "subjective social reality" from their perception of the input. It refers to the question "Is the glass half empty or half full?", used as an indicator of optimism or pessimism. To test this in animals, an individual is trained to anticipate that stimulus A, e.g. a 20 Hz tone, precedes a positive event, e.g. highly desired food is delivered when a lever is pressed by the animal. The same individual is trained to anticipate that stimulus B, e.g. a 10 Hz tone, precedes a negative event, e.g. bland food is delivered when the animal presses a lever. The animal is then tested by being played an intermediate stimulus C, e.g. a 15 Hz tone, and observing whether the animal presses the lever associated with the positive or negative reward, thereby indicating whether the animal is in a positive or negative mood. This might be influenced by, for example, the type of housing the animal is kept in.

Using this approach, it has been found that rats which are subjected to either handling or playful, experimenter-administered manual stimulation (tickling) showed different responses to the intermediate stimulus: rats exposed to tickling were more optimistic. The authors stated that they had demonstrated "for the first time a link between the directly measured positive affective state and decision making under uncertainty in an animal model."

Cognitive biases have been shown in a wide range of species including rats, dogs, rhesus macaques, sheep, chicks, starlings and honeybees.

Self-medication with Psychoactive Drugs

Humans can suffer from a range of emotional or mood disorders such as depression, anxiety, fear and panic. To treat these disorders, scientists have developed a range of psychoactive drugs such as anxiolytics. Many of these drugs are developed and tested by using a range of laboratory species. It is inconsistent to argue that these drugs are effective in treating human emotions whilst denying the experience of these emotions in the laboratory animals on which they have been developed and tested.

Standard laboratory cages prevent mice from performing several natural behaviours for which they are highly motivated. As a consequence, laboratory mice sometimes develop abnormal behaviours indicative of emotional disorders such as depression and anxiety. To improve welfare, these cages are sometimes enriched with items such as nesting material, shelters and running wheels. Sherwin and Ollson tested whether such enrichment influenced the consumption of Midazolam, a drug widely used to treat anxiety in humans. Mice in standard cages, standard cages but with unpredictable husbandry, or enriched cages, were given a choice of drinking either non-drugged water or a solution of the Midazolam. Mice in the standard and unpredictable cages drank a greater proportion of the anxiolytic solution than mice from enriched cages, indicating that mice from the standard and unpredictable laboratory caging may have been experiencing greater anxiety than mice from the enriched cages.

Spindle Neurons

Spindle neurons are specialised cells found in three very restricted regions of the human brain – the anterior cingulate cortex, the frontoinsular cortex and the dorsolateral prefrontal cortex. The first two of these areas regulate emotional functions such as empathy, speech, intuition, rapid "gut reactions" and social organization in humans. Spindle neurons are also found in the brains of humpback whales, fin whales, killer whales, sperm whales, bottlenose dolphin, Risso's dolphin, beluga whales, and the African and Asian elephants.

Whales have spindle cells in greater numbers and are maintained for twice as long as humans. The exact function of spindle cells in whale brains is still not understood, but Hof and Van Der Gucht believe that they act as some sort of "high-speed connections that fast-track information to and from other parts of the cortex". They compared them to express trains that bypass unnecessary connections, enabling organisms to instantly process and act on emotional cues during complex social interactions. However, Hof and Van Der Gucht clarify that they do not know the nature of such feelings in these animals and that we cannot just apply what we see in great apes or ourselves to whales. They believe that more work is needed to know whether emotions are the same for humans and whales.

Vocalisations

From at least the time of Darwin, it has been known that chimpanzees and other great apes also perform a laugh-like vocalization.

Research with rats has revealed that under particular conditions, they emit 50-kHz ultrasonic vocalisations (USV) which have been postulated to reflect a positive affective state (emotion) analogous to primitive human joy; these calls have been termed "laughter". The 50 kHz USVs in rats are uniquely elevated by hedonic stimuli—such as tickling, rewarding electrical brain stimulation, amphetamine injections, mating, play, and aggression—and are suppressed by aversive stimuli. Of all manipulations that elicit 50 kHz chirps in rats, tickling by humans elicits the highest rate of these calls.

Some vocalizations of domestic cats, such as purring, are well-known to be produced in situations of positive valence, such as mother kitten interactions, contacts with familiar partner, or during tactile stimulation with inanimate objects as when rolling and rubbing. Therefore, purring can be generally considered as an indicator of "pleasure" in cats.

Low pitched bleating in sheep has been associated with some positive-valence situations, as they are produced by males as an estrus female is approaching or by lactating mothers while licking and nursing their lambs.

Criticism

The argument that animals experience emotions is sometimes rejected due to a lack of evidence, and those who don't believe in the idea of animal intelligence, often argue that anthropomorphism plays a role in individuals' perspectives. Those who reject that animals have the capacity to experience emotion do so mainly by referring to inconsistencies in studies that have endorsed the belief emotions exist. Having no linguistic means to communicate emotion beyond behavioral response interpretation, the difficulty of providing an account of emotion in animals relies heavily on interpretive experimentation, that relies on results from human subjects.

Some people oppose the concept of animal emotions and suggest that emotions aren't universal, including in humans. If emotions are not universal, this indicates that there is not a phylogenetic relationship between human and non-human emotion. The relationship drawn by proponents of animal emotion, then, would be merely a suggestion of mechanistic features that promote adaptivity, but lack the complexity of human emotional constructs. Thus, a social life-style may play a role in the process of basic emotions developing into more complex emotions.

Darwin concluded, through a survey, that humans share universal emotive expressions and suggested that animals likely share in these to some degree.Social constructionists disregard the concept that emotions are universal. Others hold an intermediate stance, suggesting that basic emotional expressions and emotion are universal but the intricacies are developed culturally. A study by Elfenbein and Ambady indicated that individuals within a particular culture are better at recognising other cultural members' emotions.

Examples

Primates

Primates, in particular great apes, are candidates for being able to experience empathy and theory of mind. Great apes have complex social systems; young apes and their mothers have strong bonds of attachment and when a baby chimpanzee or gorilla dies, the mother will not uncommonly carry the body around for several days. Jane Goodall has described chimpanzees as exhibiting mournful behavior. Koko, a gorilla trained to use sign language, was reported to have expressed vocalisations indicating sadness after the death of her pet cat, All Ball.

Beyond such anecdotal evidence, support for empathetic reactions has come from experimental studies of rhesus macaques. Macaques refused to pull a chain that delivered food to themselves if doing so also caused a companion to receive an electric shock. This inhibition of hurting another conspecific was more pronounced between familiar than unfamiliar macaques, a finding similar to that of empathy in humans.

Furthermore, there has been research on consolation behavior in chimpanzees. De Waal and Aureli found that third-party contacts attempt to relieve the distress of contact participants by consoling (e.g. making contact, embracing, grooming) recipients of aggression, especially those that have experienced more intense aggression. Researchers were unable to replicate these results using the same observation protocol in studies of monkeys, demonstrating a possible difference in empathy between monkeys and apes.

Other studies have examined emotional processing in the great apes. Specifically, chimpanzees were shown video clips of emotionally charged scenes, such as a detested veterinary procedure or a favorite food, and then were required to match these scenes with one of two species-specific facial expressions: "happy" (a play-face) or "sad" (a teeth-baring expression seen in frustration or after defeat). The chimpanzees correctly matched the clips to the facial expressions that shared their meaning, demonstrating that they understand the emotional significance of their facial expressions. Measures of peripheral skin temperature also indicated that the video clips emotionally affected the chimpanzees.

Rodents

In 1998, Jaak Panksepp proposed that all mammalian species are equipped with brains capable of generating emotional experiences. Subsequent work examined studies on rodents to provide foundational support for this claim. One of these studies examined whether rats would work to alleviate the distress of a conspecific. Rats were trained to press a lever to avoid the delivery of an electric shock, signaled by a visual cue, to a conspecific. They were then tested in a situation in which either a conspecific or a Styrofoam block was hoisted into the air and could be lowered by pressing a lever. Rats that had previous experience with conspecific distress demonstrated greater than ten-fold more responses to lower a distressed conspecific compared to rats in the control group, while those who had never experienced conspecific distress expressed greater than three-fold more responses to lower a distressed conspecific relative to the control group. This suggests that rats will actively work to reduce the distress of a conspecific, a phenomenon related to empathy. Comparable results have also been found in similar experiments designed for monkeys.

Langford et al. examined empathy in rodents using an approach based in neuroscience. They reported that (1) if two mice experienced pain together, they expressed greater levels of pain-related behavior than if pain was experienced individually, (2) if experiencing different levels of pain together, the behavior of each mouse was modulated by the level of pain experienced by its social partner, and (3) sensitivity to a noxious stimulus was experienced to the same degree by the mouse observing a conspecific in pain as it was by the mouse directly experiencing the painful stimulus. The authors suggest this responsiveness to the pain of others demonstrated by mice is indicative of emotional contagion, a phenomenon associated with empathy, which has also been reported in pigs. One behaviour associated with fear in rats is freezing. If female rats experience electric shocks to the feet and then witness another rat experiencing similar footshocks, they freeze more than females without any experience of the shocks. This suggests empathy in experienced rats witnessing another individual being shocked. Furthermore, the demonstrator's behaviour was changed by the behaviour of the witness; demonstrators froze more following footshocks if their witness froze more creating an empathy loop.

Several studies have also shown rodents can respond to a conditioned stimulus that has been associated with the distress of a conspecific, as if it were paired with the direct experience of an unconditioned stimulus. These studies suggest that rodents are capable of shared affect, a concept critical to empathy.

Horses

Although not direct evidence that horses experience emotions, a 2016 study showed that domestic horses react differently to seeing photographs of positive (happy) or negative (angry) human facial expressions. When viewing angry faces, horses look more with their left eye which is associated with perceiving negative stimuli. Their heart rate also increases more quickly and they show more stress-related behaviours. One rider wrote, 'Experienced riders and trainers can learn to read the subtle moods of individual horses according to wisdom passed down from one horseman to the next, but also from years of trial-and-error. I suffered many bruised toes and nipped fingers before I could detect a curious swivel of the ears, irritated flick of the tail, or concerned crinkle above a long-lashed eye.' This suggests that horses have emotions and display them physically but it isn't concrete evidence.

Birds

Marc Bekoff reported accounts of animal behaviour which he believed was evidence of animals being able to experience emotions in his book *The Emotional Lives of Animals*. The following is an excerpt from his book:

> "A few years ago my friend Rod and I were riding our bicycles around Boulder, Colorado, when we witnessed a very interesting encounter among five magpies. Magpies are corvids, a very intelligent family of birds. One magpie had obviously been hit by a car and was laying dead on the side of the road. The four other magpies were standing around him. One approached the corpse, gently pecked at it-just as an elephant noses the carcass of another elephant- and stepped back. Another magpie did the same thing. Next, one of the magpies flew off, brought back some grass, and laid it by the corpse. Another magpie did the same. Then, all four magpies stood vigil for a few seconds and one by one flew off."

Bystander affiliation is believed to represent an expression of empathy in which the bystander tries to console a conflict victim and alleviate their distress. There is evidence for bystander affiliation in ravens (e.g. contact sitting, preening, or beak-to-beak or beak-to-body touching) and also for solicited bystander affiliation, in which there is post-conflict affiliation from the victim to the bystander. This indicates that ravens may be sensitive to the emotions of others, however, relationship value plays an important role in the prevalence and function of these post-conflict interactions.

The capacity of domestic hens to experience empathy has been studied. Mother hens show one of the essential underpinning attributes of empathy: the ability to be affected by, and share, the emotional state of their distressed chicks. However, evidence for empathy between familiar adult hens has not yet been found.

Dogs

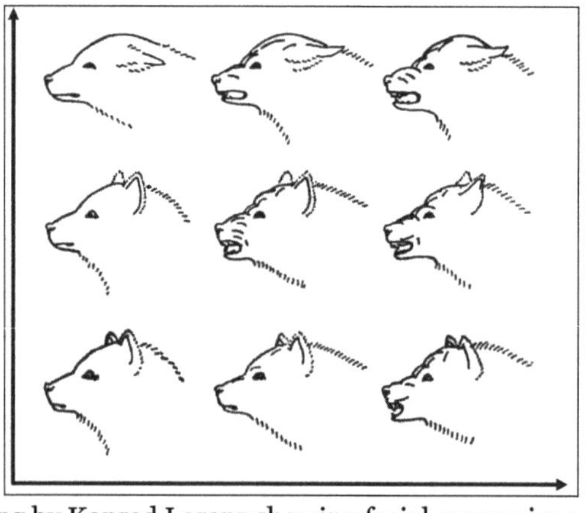

A drawing by Konrad Lorenz showing facial expressions of a dog

Some research indicates that domestic dogs may experience negative emotions in a similar manner to humans, including the equivalent of certain chronic and acute psychological conditions. Much of this is from studies by Martin Seligman on the theory of learned helplessness as an extension of his interest in depression:

A dog that had earlier been repeatedly conditioned to associate an audible stimulus with inescapable electric shocks did not subsequently try to escape the electric shocks after the warning was presented, even though all the dog would have had to do is jump over a low divider within ten seconds. The dog didn't even try to avoid the "aversive stimulus"; it had previously "learned" that nothing it did would reduce the probability of it receiving a shock. A follow-up experiment involved three dogs affixed in harnesses, including one that received shocks of identical intensity and duration to the others, but the lever which would otherwise have allowed the dog a degree of control was left disconnected and didn't do anything. The first two dogs quickly recovered from the experience, but the third dog suffered chronic symptoms of clinical depression as a result of this perceived helplessness.

A further series of experiments showed that, similar to humans, under conditions of long-term intense psychological stress, around one third of dogs do not develop learned helplessness or long term depression. Instead these animals somehow managed to find a way to handle the unpleasant situation in spite of their past experience. The corresponding characteristic in humans has been found to correlate highly with an explanatory style and optimistic attitude that views the situation as *other than* personal, pervasive, or permanent.

Since these studies, symptoms analogous to clinical depression, neurosis, and other psychological conditions have also been accepted as being within the scope of emotion in domestic dogs. The postures of dogs may indicate their emotional state. In some instances, the recognition of specific postures and behaviors can be learned.

Psychology research has shown that when humans gaze at the face of another human, the gaze is not symmetrical; the gaze instinctively moves to the right side of the face to obtain information about their emotions and state. Research at the University of Lincoln shows that dogs share this instinct when meeting a human, and only when meeting a human (i.e. not other animals or other dogs). They are the only non-primate species known to share this instinct.

The existence and nature of personality traits in dogs have been studied (15,329 dogs of 164 different breeds). Five consistent and stable "narrow traits" were identified, described as playfulness, curiosity/fearlessness, chase-proneness, sociability and aggressiveness. A further higher order axis for shyness–boldness was also identified.

Dogs presented with images of either human or dog faces with different emotional states (happy/playful or angry/aggressive) paired with a single vocalization (voices or barks) from the same individual with either a positive or negative emotional state or brown noise. Dogs look longer at the face whose expression is congruent to the emotional state of the vocalization, for both other dogs and humans. This is an ability previously known only in humans. The behavior of a dog can not always be an indication its friendliness. This is because when a dog wags its tail, most people interpret this as the dog expressing happiness and friendliness. Though indeed tail wagging can express these positive emotions, tail wagging is also an indication of fear, insecurity, challenging of dominance, establishing social relationships or a warning that the dog may bite.

Some researchers are beginning to investigate the question of whether dogs have emotions with the help of magnetic resonance imaging.

Cats

It has been postulated that domestic cats can learn to manipulate their owners through vocalizations that are similar to the cries of human babies. Some cats learn to add a purr to the vocalization, which makes it less harmonious and more dissonant to humans, and therefore harder to ignore. Individual cats learn to make these vocalizations through trial-and-error; when a particular vocalization elicits a positive response from a human, the probability increases that the cat will use that vocalization in the future.

Growling can be an expression of annoyance or fear, similar to humans. When annoyed or angry, a cat wriggles and thumps its tail much more vigorously than when in a contented state. In larger felids such as lions, what appears to be irritating to them varies between individuals. A male lion may let his cubs play with his mane or tail, or he may hiss and hit them with his paws. Domestic male cats also have variable attitudes towards their family members, for example, older male siblings tend not to go near younger or new siblings and may even show hostility toward them.

Honeybees

Honeybees become pessimistic after being shaken

Honeybees ("Apis mellifera carnica") were trained to extend their proboscis to a two-component odour mixture (CS+) predicting a reward (e.g., 1.00 or 2.00 M sucrose) and to withhold their proboscis from another mixture (CS−) predicting either punishment or a less valuable reward (e.g., 0.01 M quinine solution or 0.3 M sucrose). Immediately after training, half of the honeybees were subjected to vigorous shaking for 60 s to simulate the state produced by a predatory attack on a concealed colony. This shaking reduced levels of octopamine, dopamine, and serotonin in the hemolymph of a separate group of honeybees at a time point corresponding to when the cognitive bias tests were performed. In honeybees, octopamine is the local neurotransmitter that functions during reward learning, whereas dopamine mediates the ability to learn to associate odours with quinine punishment. If flies are fed serotonin, they are more aggressive; flies depleted of serotonin still exhibit aggression, but they do so much less frequently.

Within 5 minutes of the shaking, all the trained bees began a sequence of unreinforced test trials with five odour stimuli presented in a random order for each bee: the CS+, the CS−, and three novel odours composed of ratios intermediate between the two learned mixtures. Shaken honeybees were more likely to withhold their mouthparts from the CS− and from the most similar novel odour. Therefore, agitated honeybees display an increased expectation of bad outcomes similar to a vertebrate-like emotional state. The researchers of the study stated that, "Although our results do not allow us to make any claims about the presence of negative subjective feelings in honeybees, they call into question how we identify emotions in any non-human animal. It is logically inconsistent to claim that the presence of pessimistic cognitive biases should be taken as confirmation that dogs or rats are anxious but to deny the same conclusion in the case of honeybees."

Crayfish

Crayfish naturally explore new environments but display a general preference for dark places. A 2014 study on the freshwater crayfish *Procambarus clarkii* tested their responses in a fear paradigm, the elevated plus maze in which animals choose to walk on an elevated cross which offers both aversive and preferable conditions (in this case, two arms were lit and two were dark). Crayfish which experienced an electric shock displayed enhanced fearfulness or anxiety as demonstrated by their preference for the dark arms more than the light. Furthermore, shocked crayfish had relatively higher brain serotonin concentrations coupled with elevated blood glucose, which indicates a stress response. Moreover, the crayfish calmed down when they were injected with the benzodiazepine anxiolytic, chlordiazepoxide, used to treat anxiety in humans, and they entered the dark as normal. The authors of the study concluded "stress-induced avoidance behavior in crayfish exhibits striking homologies with vertebrate anxiety."

The freshwater crayfish *Procambarus clarkii*

A follow-up study using the same species confirmed the anxiolytic effect of chlordiazepoxide, but moreover, the intensity of the anxiety-like behaviour was dependent on the intensity of the electric shock until reaching a plateau. Such a quantitative relationship between stress and anxiety is also a very common feature of human and vertebrate anxiety.

Pain in Animals

In humans, pain is a distressing feeling often caused by intense or damaging stimuli. Whether animals apart from humans also experience pain is often contentious despite being scientifically verifiable. The standard measure of pain in humans is how a person reports that pain, (for example, on a pain scale). "Pain" is defined by the International Association for the Study of Pain as "an unpleasant sensory and emotional experience associated with actual or potential tissue damage, or described in terms of such damage." Only the person experiencing the pain can know the pain's quality and intensity, and the degree of suffering. However, for non-human animals, it is harder, if even possible, to know whether an emotional experience has occurred. Therefore, this concept is often excluded in definitions of pain in animals, such as that provided by Zimmerman: "an aversive sensory experience caused by actual or potential injury that elicits protective motor and vegetative reactions, results in learned avoidance and may modify species-specific behaviour, including social behaviour." Non-human animals cannot report their feelings to language-using humans in the same manner as human communication, but observation of their behaviour provides a reasonable indication as to the extent of their pain. Just as with doctors and medics who sometimes share no common language with their patients, the indicators of pain can still be understood.

A Galapagos shark hooked by a fishing boat

According to the U.S. National Research Council Committee on Recognition and Alleviation of Pain in Laboratory Animals, pain is experienced by many animal species, including mammals and possibly all vertebrates.

The Experience of Pain

Although there are numerous definitions of pain, almost all involve two key components. First, nociception is required. This is the ability to detect noxious stimuli which evoke a reflex response that rapidly moves the entire animal, or the affected part of its body, away from the source of the stimulus. The concept of nociception does not imply any adverse, subjective "feeling" – it is a reflex action. An example in humans would be the rapid withdrawal of a finger that has touched something hot – the withdrawal occurs before any sensation of pain is actually experienced.

The second component is the experience of "pain" itself, or suffering – the internal, emotional interpretation of the nociceptive experience. Again in humans, this is when the withdrawn finger begins to hurt, moments after the withdrawal. Pain is therefore a private, emotional experience. Pain cannot be directly measured in other animals, including other humans; responses to putatively painful stimuli can be measured, but not the experience itself. To address this problem when assessing the capacity of other species to experience pain, argument-by-analogy is used. This is based on the principle that if an animal responds to a stimulus in a similar way to ourselves, it is likely to have had an analogous experience.

Reflex Response to Painful Stimuli

Nociception usually involves the transmission of a signal along nerve fibers from the site of a noxious stimulus at the periphery to the spinal cord. Although this signal is also transmitted on to the brain, a reflex response, such as flinching or withdrawal of a limb, is produced by return signals originating in the spinal cord. Thus, both physiological and behavioral responses to nociception can be detected, and no reference need be made to a conscious experience of pain. Based such criteria nociception has been observed in all major animal taxa.

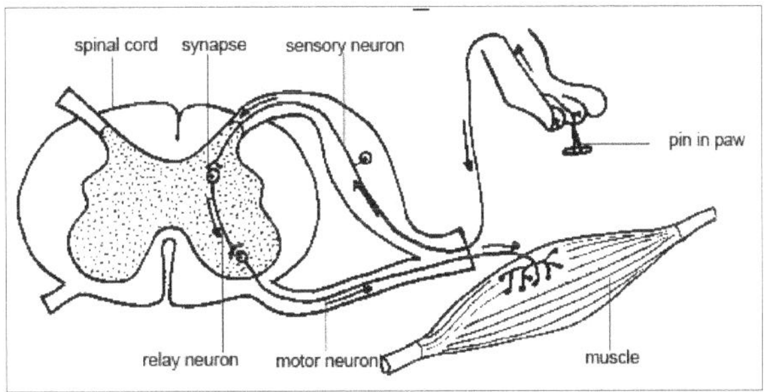

Reflex arc of a dog when its paw is stuck with a pin. The spinal cord responds to signals from receptors in the paw, producing a reflex withdrawal of the paw. This localized response does not involve brain processes that might mediate a consciousness of pain, though these might also occur

Awareness of Pain

Nerve impulses from nociceptors may reach the brain, where information about the stimulus (e.g. quality, location, and intensity), and affect (unpleasantness) are registered. Though the brain activity involved has been studied, the brain processes underlying conscious awareness are not well understood.

Adaptive Value

The adaptive value of nociception is obvious; an organism detecting a noxious stimulus immediately withdraws the limb, appendage or entire body from the noxious stimulus and thereby avoids further (potential) injury. However, a characteristic of pain (in mammals at least) is that pain can result in hyperalgesia (a heightened sensitivity to noxious stimuli) and allodynia (a heightened sensitivity to non-noxious stimuli). When this heightened sensitisation occurs, the adaptive value is less clear. First, the pain arising from the heightened sensitisation can be disproportionate to the actual tissue damage caused. Second, the heightened sensitisation may also become chronic, persisting well beyond the tissues healing. This can mean that rather than the actual tissue damage causing pain, it is the pain due to the heightened sensitisation that becomes the concern. This means the sensitisation process is sometimes termed maladaptive. It is often suggested hyperalgesia and allodynia assist organisms to protect themselves during healing, but experimental evidence to support this has been lacking.

In 2014, the adaptive value of sensitisation due to injury was tested using the predatory interactions between longfin inshore squid (*Doryteuthis pealeii*) and black sea bass (*Centropristis striata*) which are natural predators of this squid. If injured squid are targeted by a bass, they began their defensive behaviours sooner (indicated by greater alert distances and longer flight initiation distances) than uninjured squid. If anaesthetic (1% ethanol and $MgCl_2$) is administered prior to the injury, this prevents the sensitisation and blocks the behavioural effect. The authors claim this study is the first experimental evidence to support the argument that nociceptive sensitisation is actually an adaptive response to injuries.

Argument-by-analogy

To assess the capacity of other species to consciously suffer pain we resort to argument-by-analogy. That is, if an animal responds to a stimulus the way a human does, it is likely to have had an analogous experience. If we stick a pin in a chimpanzee's finger and she rapidly withdraws her hand, we use argument-by-analogy and infer that like us, she felt pain. It might be argued that consistency requires us infer, also, that a cockroach experiences conscious pain when it writhes after being stuck with a pin. The usual counter-argument is that although the physiology of consciousness is not understood, it clearly involves complex brain processes not present in relatively simple organisms. Other analogies have been pointed out. For example, when given a choice of foods, rats and chickens with clinical symptoms of pain will consume more of an analgesic-containing food than animals not in pain. Additionally, the consumption of the analgesic carprofen in lame chickens was positively correlated to the severity of lameness, and consumption resulted in an improved gait. Such anthropomorphic arguments face the criticism that physical reactions indicating pain may be neither the cause nor result of conscious states, and the approach is subject to criticism of anthropomorphic interpretation. For example, a single-celled organism such as an amoeba may writhe after being exposed to noxious stimuli despite the absence of nociception.

In Different Species

The ability to experience pain in an animal, or another human for that matter, cannot be determined directly but it may be inferred through analogous physiological and behavioral reactions. Although many animals share similar mechanisms of pain detection to those of humans, have similar areas of the brain involved in processing pain, and show similar pain behaviours, it is notoriously difficult to assess how animals actually experience pain.

Nociception

Nociceptive nerves, which preferentially detect (potential) injury-causing stimuli, have been identified in a variety of animals, including invertebrates. The medicinal leech, *Hirudo medicinalis,* and sea slug are classic model systems for studying nociception. Many other vertebrate and invertebrate animals also show nociceptive reflex responses similar to our own.

Pain

Many animals also exhibit more complex behavioural and physiological changes indicative of the ability to experience pain: they eat less food, their normal behaviour is disrupted, their social behaviour is suppressed, they may adopt unusual behaviour patterns, they may emit characteristic distress calls, experience respiratory and cardiovascular changes, as well as inflammation and release of stress hormones.

Some criteria that may indicate the potential of another species to feel pain include:

1. Has a suitable nervous system and sensory receptors.
2. Physiological changes to noxious stimuli.
3. Displays protective motor reactions that might include reduced use of an affected area such as limping, rubbing, holding or autotomy.
4. Has opioid receptors and shows reduced responses to noxious stimuli when given analgesics and local anaesthetics.

5. Shows trade-offs between stimulus avoidance and other motivational requirements.
6. Shows avoidance learning.
7. High cognitive ability and sentience.

Vertebrates

Fish

A typical human cutaneous nerve contains 83% C type trauma receptors (the type responsible for transmitting signals described by humans as excruciating pain); the same nerves in humans with congenital insensitivity to pain have only 24-28% C type receptors. The rainbow trout has about 5% C type fibres, while sharks and rays have 0%. Nevertheless, fish have been shown to have sensory neurons that are sensitive to damaging stimuli and are physiologically identical to human nociceptors. Behavioural and physiological responses to a painful event appear comparable to those seen in amphibians, birds, and mammals, and administration of an analgesic drug reduces these responses in fish.

Animal welfare advocates have raised concerns about the possible suffering of fish caused by angling. Some countries, e.g. Germany, have banned specific types of fishing, and the British RSPCA now formally prosecutes individuals who are cruel to fish.

Invertebrates

Though it has been argued that most invertebrates do not feel pain, there is some evidence that invertebrates, especially the decapod crustaceans (e.g. crabs and lobsters) and cephalopods (e.g. octopuses), exhibit behavioural and physiological reactions indicating they may have the capacity for this experience. Nociceptors have been found in nematodes, annelids and molluscs. Most insects do not possess nociceptors, one known exception being the fruit fly. In vertebrates, endogenous opioids are neurochemicals that moderate pain by interacting with opiate receptors. Opioid peptides and opiate receptors occur naturally in nematodes, molluscs, insects and crustaceans. The presence of opioids in crustaceans has been interpreted as an indication that lobsters may be able to experience pain, although it has been claimed "at present no certain conclusion can be drawn".

One suggested reason for rejecting a pain experience in invertebrates is that invertebrate brains are too small. However, brain size does not necessarily equate to complexity of function. Moreover, weight for body-weight, the cephalopod brain is in the same size bracket as the vertebrate brain, smaller than that of birds and mammals, but as big as or bigger than most fish brains.

Since September 2010, all cephalopods being used for scientific purposes in the EU are protected by EU Directive 2010/63/EU which states "there is scientific evidence of their [cephalopods] ability to experience pain, suffering, distress and lasting harm. In the UK, animal protection legislation means that cephalopods used for scientific purposes must be killed humanely, according to prescribed methods (known as "Schedule 1 methods of euthanasia") known to minimise suffering.

In Medicine and Research

Veterinary Medicine

Veterinary medicine uses, for actual or potential animal pain, the same analgesics and anesthetics as used in humans.

Dolorimetry

Dolorimetry (*dolor*: Latin: pain, grief) is the measurement of the pain response in animals, including humans. It is practiced occasionally in medicine, as a diagnostic tool, and is regularly used in research into the basic science of pain, and in testing the efficacy of analgesics. Non-human animal pain measurement techniques include the paw pressure test, tail flick test, hot plate test and grimace scales.

Laboratory Animals

Animals are kept in laboratories for a wide range of reasons, some of which may involve pain, suffering or distress, whilst others (e.g. many of those involved in breeding) will not. The extent to which animal testing causes pain and suffering in laboratory animals is the subject of much debate. Marian Stamp Dawkins defines "suffering" in laboratory animals as the experience of one of "a wide range of extremely unpleasant subjective (mental) states." The U.S. National Research Council has published guidelines on the care and use of laboratory animals, as well as a report on recognizing and alleviating pain in vertebrates. The United States Department of Agriculture defines a "painful procedure" in an animal study as one that would "reasonably be expected to cause more than slight or momentary pain or distress in a human being to which that procedure was applied." Some critics argue that, paradoxically, researchers raised in the era of increased awareness of animal welfare may be inclined to deny that animals are in pain simply because they do not want to see themselves as people who inflict it. PETA however argues that there is no doubt about animals in laboratories being inflicted with pain. In the UK, animal research likely to cause "pain, suffering, distress or lasting harm" is regulated by the Animals (Scientific Procedures) Act 1986 and research with the potential to cause pain is regulated by the Animal Welfare Act of 1966 in the US.

In the U.S., researchers are not required to provide laboratory animals with pain relief if the administration of such drugs would interfere with their experiment. Laboratory animal veterinarian Larry Carbone writes, "Without question, present public policy allows humans to cause laboratory animals unalleviated pain. The AWA, the *Guide for the Care and Use of Laboratory Animals*, and current Public Health Service policy all allow for the conduct of what are often called "Category E" studies – experiments in which animals are expected to undergo significant pain or distress that will be left untreated because treatments for pain would be expected to interfere with the experiment."

Severity Scales

Eleven countries have national classification systems of pain and suffering experienced by animals used in research: Australia, Canada, Finland, Germany, The Republic of Ireland, The Netherlands,

New Zealand, Poland, Sweden, Switzerland, and the UK. The US also has a mandated national scientific animal-use classification system, but it is markedly different from other countries in that it reports on whether pain-relieving drugs were required and/or used. The first severity scales were implemented in 1986 by Finland and the UK. The number of severity categories ranges between 3 (Sweden and Finland) and 9 (Australia). In the UK, research projects are classified as "mild", "moderate", and "substantial" in terms of the suffering the researchers conducting the study say they may cause; a fourth category of "unclassified" means the animal was anesthetized and killed without recovering consciousness. It should be remembered that in the UK system, many research projects (e.g. transgenic breeding, feeding distasteful food) will require a license under the Animals (Scientific Procedures) Act 1986, but may cause little or no pain or suffering. In December 2001, 39 percent (1, 296) of project licenses in force were classified as "mild", 55 percent (1, 811) as "moderate", two percent (63) as "substantial", and 4 percent (139) as "unclassified". In 2009, of the project licenses issued, 35 percent (187) were classified as "mild", 61 percent (330) as "moderate", 2 percent (13) as "severe" and 2 percent (11) as unclassified.

In the US, the *Guide for the Care and Use of Laboratory Animals* defines the parameters for animal testing regulations. It states, "The ability to experience and respond to pain is widespread in the animal kingdom...Pain is a stressor and, if not relieved, can lead to unacceptable levels of stress and distress in animals. " The *Guide* states that the ability to recognize the symptoms of pain in different species is essential for the people caring for and using animals. Accordingly, all issues of animal pain and distress, and their potential treatment with analgesia and anesthesia, are required regulatory issues for animal protocol approval.

Tool use by Animals

A crab-eating macaque using a stone

Tool use by animals is a phenomenon in which an animal uses any kind of tool in order to achieve a goal such as acquiring food and water, grooming, defense, recreation or construction. Originally thought to be a skill only possessed by humans, some tool use requires a sophisticated level of cognition. There is considerable discussion about the definition of what constitutes a tool and therefore which behaviours can be considered as true examples of tool use. A wide range of animals are considered to use tools including mammals, birds, fish, cephalopods and insects.

Primates are well known for using tools for hunting or gathering food and water, cover for rain, and self-defence. Chimpanzees have been the object of study, most famously by Jane Goodall, since these animals are more-often kept in captivity than other primates and are closely related to humans. Tool-use in other primates are lesser-known as many of them are mainly observed in the wild. Many famous researchers, such as Charles Darwin in his book *The Descent of Man*, mentioned tool-use in monkeys (such as baboons). Both wild and captive elephants are known to create tools using their trunk and feet, mainly for swatting flies, scratching, plugging water-holes (so the water doesn't evaporate), and reaching food that is out of reach. A group of dolphins in Shark Bay use sponges to protect their beak while foraging. Sea otters will dislodge food from rocks (such as abalone) and break open shellfish. Carnivores (of the order Carnivora) can use tools to trap prey or break open the shells of prey, as well as for scratching.

Corvids (crows, ravens and rooks) are well known for their large brains (among birds) and subsequent tool use. They mainly manufacture probes out of twigs and wood (and sometimes metal wire) to catch or impale larvae. Crows are among the only animals that create their own toys. Tool use in other birds is best exemplified in nest intricacy. Warblers manufacture 'pouches' to make their nests in. Some birds, such as weaver birds build complex nests. Finches and woodpeckers may insert twigs into trees in order to catch or impale larvae. Parrots may use tools to wedge nuts so that they may crack it open (using a tool) without launching it away. Some birds take advantage of human activity, such as some species of gulls which drop shellfish in front of cars to crack them open.

Several species of fish use tools to crack open shellfish, extract food that is out of reach, cleaning an area (for nesting), and hunting. Octopuses gather coconut shells and create a shelter. They may also construct a fence using rocks.

Definitions and Terminology

The key to identifying tool use is defining what constitutes a tool. Researchers of animal behavior have arrived at different formulations.

In 1980, Beck published a widely used definition of tool use. This has been modified to:

The external employment of an unattached or manipulable attached environmental object to alter more efficiently the form, position, or condition of another object, another organism, or the user itself, when the user holds and directly manipulates the tool during or prior to use and is responsible for the proper and effective orientation of the tool.

Other, briefer definitions have been proposed:

An object carried or maintained for future use.

—Finn, Tregenza, and Norman, 2009.

The use of physical objects other than the animal's own body or appendages as a means to extend the physical influence realized by the animal.

—Jones and Kamil, 1973

An object that has been modified to fit a purpose ... [or] An inanimate object that one uses or modifies in some way to cause a change in the environment, thereby facilitating one's achievement of a target goal.

—Hauser, 2000

Others, for example Lawick-Goodall, distinguish between "tool use" and "object use".

Different terms have been given to the tool according to whether the tool is altered by the animal. If the "tool" is not held or manipulated by the animal in any way, such as an immobile anvil, objects in a bowerbird's bower, or a bird using bread as bait to catch fish, it is sometimes referred to as a "proto-tool". Several studies in primates and birds have found that tool use is correlated with an enlargement of the brain as a whole or of particular regions. For example, true tool-using birds have relatively larger brains than proto-tool users.

When an animal uses a tool that acts on another tool, this has been termed use of a "meta-tool". For example, New Caledonian crows will spontaneously use a short tool to obtain an otherwise inaccessible longer tool that then allows them to extract food from a hole. Similarly, bearded capuchin monkeys will use smaller stones to loosen bigger quartz pebbles embedded in conglomerate rock, which they subsequently use as tools.

Rarely, animals may use one tool followed by another, for example, bearded capuchins use stones and sticks, or two stones. This is called "associative", "secondary" or "sequential" tool use.

Some animals use other individuals in a way which could be interpreted as tool use, for example, ants crossing water over a bridge of other ants, or weaver ants using conspecifics to glue leaves together. These have been termed "social tools".

Borderline Examples

Play: Play has been defined as "activity having no immediate benefits and structurally including repetitive or exaggerated actions that may be out of sequence or disordered". When play is discussed in relation to manipulating objects, it is often used in association with the word "tool". Some birds, notably crows, parrots and birds of prey, "play" with objects, many of them playing in flight with such items as stones, sticks and leaves, by letting them go and catching them again before they reach the ground. A few species repeatedly drop stones, apparently for the enjoyment of the sound effects. Many other species of animals, both avian and non-avian, play with objects in a similar manner.

Fixed "devices": The impaling of prey on thorns by many of the shrikes (*Laniidae*) is well known. Several other birds may use spines or forked sticks to anchor a carcass while they flay it with the bill. It has been concluded that "This is an example of a fixed device which serves as an extension of the body, in this case, talons" and is thus a true form of tool use. On the other hand, the use of fixed skewers may not be true tool-use because the thorn (or whatever) is not manipulated by the bird. Leopards perform a similar behaviour by dragging carcasses up trees and caching them in the forks of branches.

Use of bait: Several species of bird, including herons such as the striated heron (*Butorides striatus*), will place bread in water to attract fish. Whether this is tool use is disputed because the bread is not manipulated or held by the bird.

Learning and Cognition

Tool use by animals may indicate different levels of learning and cognition. For some animals, tool use is largely instinctive and inflexible. For example, the woodpecker finch of the Galápagos Islands use twigs or spines as an essential and regular part of its foraging behaviour, but these behaviours are often quite inflexible and are not applied effectively in different situations. Other tool use, e.g. chimpanzees using twigs to "fish" for termites, may be developed by watching others use tools and may even be a true example of animal teaching. Tools may even be used in solving puzzles in which the animal appears to experience a "Eureka moment".

Primates

A gorilla using a stick possibly to gauge the depth of water

Tool use has been reported many times in both wild and captive primates, particularly the great apes. The use of tools by primates is varied and includes hunting (mammals, invertebrates, fish), collecting honey, processing food (nuts, fruits, vegetables and seeds), collecting water, weapons and shelter.

In 1960, Jane Goodall observed a chimpanzee poking pieces of grass into a termite mound and then raising the grass to his mouth. After he left, Goodall approached the mound and repeated the behaviour because she was unsure what the chimpanzee was doing. She found that the termites bit onto the grass with their jaws. The chimpanzee had been using the grass as a tool to "fish" or "dip" for termites.

Tool manufacture is much rarer than simple tool use and probably represents higher cognitive functioning. Soon after her initial discovery of tool use, Goodall observed other chimpanzees picking up leafy twigs, stripping off the leaves and using the stems to fish for insects. This change of a leafy twig into a tool was a major discovery. Prior to this, scientists thought that only humans manufactured and used tools, and that this ability was what separated humans from other animals. In 1990, it was claimed the only primate to manufacture tools in the wild was the chimpanzee. However, since then, several primates have been reported as tool makers in the wild.

Both bonobos and chimpanzees have been observed making "sponges" out of leaves and moss that suck up water and using these for grooming. Sumatran orangutans will take a live branch, remove twigs and leaves and sometimes the bark, before fraying or flattening the tip for use on ants or bees. In the wild, mandrills have been observed to clean their ears with modified tools. Scientists filmed a large male mandrill at Chester Zoo (UK) stripping down a twig, apparently to make it narrower, and then using the modified stick to scrape dirt from underneath his toenails. Captive gorillas have made a variety of tools.

Chimpanzees and Bonobos

Common chimpanzees (*Pan troglodytes*) are sophisticated tool users with behaviors including cracking nuts with stone tools and fishing for ants or termites with sticks. There are more limited reports of the closely related bonobo (*Pan paniscus*) using tools in the wild; it has been claimed they rarely use tools in the wild although they use tools as readily as chimpanzees when in captivity, It has been reported that females of both chimpanzees and bonobos use tools more avidly than males. Wild chimpanzees predominantly use tools in the context of food acquisition, while wild bonobos appear to use tools mainly for personal care (cleaning, protection from rain) and social purposes. Wild bonobos have been observed using leaves as cover for rain, or the use of branches in social displays.

Hunting

Research in 2007 showed that common chimpanzees sharpen sticks to use as weapons when hunting mammals. This is considered the first evidence of systematic use of weapons in a species other than humans. Researchers documented 22 occasions when wild chimpanzees on a savanna in Senegal fashioned sticks into "spears" to hunt lesser bushbabies (*Galago senegalensis*). In each case, a chimpanzee modified a branch by breaking off one or two ends and, frequently using its teeth, sharpened the stick. The tools, on average, were about 60 cm (24 in) long and 1.1 cm (0.4 in) in circumference. The chimpanzee then jabbed the spear into hollows in tree trunks where bushbabies sleep. There was a single case in which a chimpanzee successfully extracted a bushbaby with the tool. It has been suggested that the word "spear" is an overstatement that makes the chimpanzees seem too much like early humans, and that the term "bludgeon" is more accurate, since the point of the tool may not be particularly sharp. This behaviour was seen more frequently in females, particularly adolescent females, and young chimps in general, than in adult males.

Chimpanzees often eat the marrow of long bones of colobus monkeys with the help of small sticks, after opening the ends of the bones with their teeth. A juvenile female was observed to eat small parts of the brain of an intact skull that she could not break open by inserting a small stick through the foramen magnum. On another occasion, an adult female used three sticks to clean the orbits of a colobus monkey skull after she had just eaten the eyes.

A Bonobo Fishing for Termites

In Gombe National Park in 1960, Jane Goodall observed a chimpanzee, David Greybeard, poking pieces of grass into a termite mound and then raising the grass to his mouth. After he left, Goodall approached the mound and repeated the behaviour because she was unsure what David was doing. She found that the termites bit onto the grass with their jaws. David had been using the grass as a tool to "fish" or "dip" for termites. Soon after this initial discovery of tool use, Goodall observed

David and other chimpanzees picking up leafy twigs, stripping off the leaves, and using the stems to fish for insects. This modification of a leafy twig into a tool was a major discovery: previously, scientists thought that only humans made and used tools, and that this was what separated humans from other animals.

Other studies of the Gombe chimps show that young females and males learn to fish for termites differently. Female chimps learn to fish for termites earlier and better than the young males. Females also spend more time fishing while at the mounds with their mothers—males spend more time playing. When they are adults, females need more termite protein because with young to care for, they cannot hunt the way males can.

Populations differ in the prevalence of tool use for fishing for invertebrates. Chimpanzees in the Tai National Park only sometimes use tools, whereas Gombe chimpanzees rely almost exclusively on tools for their intake of driver ants. This may be due to difference in the rewards gained by tool use: Gombe chimpanzees collect 760 ants/min compared to 180 ants/min for the Tai chimpanzees.

Some chimpanzees use tools to hunt large bees (*Xylocopa* sp.) which make nests in dead branches on the ground or in trees. To get to the grubs and the honey, the chimpanzee first tests for the presence of adults by probing the nest entrance with a stick. If present, adult bees block the entrance with their abdomens, ready to sting. The chimpanzee then disables them with the stick to make them fall out and eats them rapidly. Afterwards, the chimpanzee opens the branch with its teeth to obtain the grubs and the honey.

Chimpanzees have even been observed using two tools: a stick to dig into an ant nest and a "brush" made from grass stems with their teeth to collect the ants.

Collecting Honey

Honey of four bee species is eaten by chimpanzees. Groups of chimpanzees fish with sticks for the honey after having tried to remove what they can with their hands. They usually extract with their hands honeycombs from undisturbed hives of honey bees and run away from the bees to quietly eat their catch. In contrast, hives that have already been disturbed, either through the falling of the tree or because of the intervention of other predators, are cleaned of the remaining honey with fishing tools.

Processing Food

A chimpanzee gathering food with a stick

Tai chimpanzees crack open nuts with rocks, but there is no record of Gombe chimpanzees using rocks in this way. After opening nuts by pounding with a hammer, parts of the kernels may be too difficult to reach with the teeth or fingernails, and some individuals use sticks to remove these remains, instead of pounding the nut further with the hammer as other individuals do: a relatively rare combination of using two different tools. Hammers for opening nuts may be either wood or stone.

Chimpanzees in the Nimba Mountains of Guinea, Africa, use both stone and wooden cleavers, as well as stone anvils, to chop up and reduce *Treculia* fruits into smaller bite-sized portions. These fruits, which can be the size of a volleyball and weigh up to 8.5 kg, are hard and fibrous. But, despite lacking a hard outer shell, they are too large for a chimpanzee to get its jaws around and bite into. Instead, the chimpanzees use a range of tools to chop them into smaller pieces. This is the first account of chimpanzees using a pounding tool technology to break down large food items into bite-sized chunks rather than just extracting it from other unobtainable sources such as baobab nuts. It is also the first time wild chimpanzees have been found to use two distinct types of percussive technology, i.e. movable cleavers against a non-movable anvil, to achieve the same goal. Neighbouring chimpanzees in the nearby region of Seringbara do not process their food in this way, indicating how tool use among apes is culturally learnt.

Collecting Water

When chimpanzees cannot reach water that has formed in hollows high up inside trees, they have been observed taking a handful of leaves, chewing them, and dipping this "sponge" into the pool to suck out the water. Both bonobos and chimpanzees have also been observed making "sponges" out of leaves and moss that suck up water and are used as grooming tools.

Orangutans

Orangutans (genus *Pongo*) were first observed using tools in the wild in 1994 in the northwest corner of Sumatra. As with the chimpanzees, orangutans use tools made from branches and leaves to scratch, scrape, wipe, sponge, swat, fan, hook, probe, scoop, pry, chisel, hammer, cover, cushion and amplify. They will break off a tree branch that is about 30 cm long, snap off the twigs, fray one end and then use the stick to dig in tree holes for termites. Sumatran orangutans use a variety of tools—up to 54 types for extracting insects or honey, and as many as 20 types for opening or preparing fruits such as the hard to access *Neesia Malayana*. They also use an 'autoerotic tool'—a stick which they use to stimulate the genitals and masturbate (both male and female). In parts of Borneo, orangutans use handfuls of leaves as napkins to wipe their chins while orangutans in parts of Sumatra use leaves as gloves, helping them handle spiny fruits and branches, or as seat cushions in spiny trees. There have been reports that individuals in both captivity and in the wild use tools held between the lips or teeth, rather than in the hands. In captivity, orangutans have been taught to chip stone handaxes.

Orangutans living in Borneo scavenge fish that wash up along the shore and scoop catfish out of small ponds for fresh meals. Over two years, anthropologist Anne Russon saw several animals on these forested islands learn on their own to jab at catfish with sticks, so that the panicked prey would flop out of ponds and into the orangutan's waiting hands. Although orangutans usually fished alone, Russon observed pairs of apes catching catfish on a few occasions. On the island of Kaja in Borneo, a male orangutan was observed using a pole apparently trying to spear or bludgeon fish. This individual had seen humans fishing with spears. Although not successful, he was later able to improvise by using the pole to catch fish already trapped in the locals' fishing lines.

Sumatran orangutans use sticks to acquire seeds from a particular fruit. When the fruit of the *Neesia* tree ripens, its hard, ridged husk softens until it falls open. Inside are seeds that are highly desirable to the orangutans, but they are surrounded by fibreglass-like hairs that are painful if eaten. A *Neesia*-eating orangutan will select a 12 cm stick, strip off the bark, and then carefully collect the hairs with it. Once the fruit is safe, the ape will eat the seeds using the stick or its fingers. Sumatran orangutans will use a stick to poke a bees' nest wall, move it around and catch the honey.

Orangutans have been observed using sticks to apparently measure the depth of water. It has been reported that orangutans use tools for a wide range of purposes including using leaves as protective gloves or napkins, using leafy branches to swat insects or gather water, and building sun or rain covers above the nests used for resting. It has been reported that a Sumatran orangutan used a large leaf as an umbrella in a tropical rainstorm.

Orangutans produce an alarm call known as a "kiss squeak" when they encounter a predator like a snake or a human. Sometimes, orangutans will strip leaves from a branch and hold them in front of their mouth when making the sound. It has been found this lowers the maximum frequency of the sound i.e. makes it deeper, and in addition, smaller orangutans are more likely to use the leaves. It has been suggested they use the leaves to make themselves sound bigger than they really are, the first documented case of an animal using a tool to manipulate sound.

Gorillas

A gorilla pushing a stick into the ground and using it as a stabilizing stick while dredging aquatic herbs

There are few reports of gorillas using tools in the wild. Western lowland gorillas (*Gorilla gorilla gorilla*) have been observed using sticks to apparently measure the depth of water and as "walking sticks" to support their posture when crossing deeper water. An adult female used a detached trunk from a small shrub as a stabilizer during food gathering, and another used a log as a bridge. One possible explanation for the absence of observed tool use in wild gorillas is that they are less dependent on foraging techniques that require the use of tools, since they exploit food resources differently from chimpanzees. Whereas chimpanzees and orangutans feeding involves tools such as hammers to crack open nuts and sticks to fish for termites, gorillas access these foods by breaking nuts with their teeth and smashing termite mounds with their hands.

Captive western lowland gorillas have been observed to threaten each other with sticks and larger pieces of wood, while others use sticks for hygienic purposes. Some females have attempted to use logs as ladders. In another group of captive gorillas, several individuals were observed throwing sticks and branches into a tree, apparently to knock down leaves and seeds. Gorillas at Prague zoo have used tools in several ways, including using wood wool as "slippers" when walking on the snow or to cross a wet section of the floor.

Monkeys

Tool use has been observed in capuchin monkeys both in captivity and in their natural environments. In a captive environment, capuchins readily insert a stick into a tube containing viscous food that clings to the stick, which they then extract and lick. Capuchins also use a stick to push food from the center of a tube retrieving the food when it reaches the far end, and as a rake to sweep objects or food toward themselves. The black-striped capuchin (*Sapajus libidinosus*) was the first non-ape primate for which tool use was documented in the wild; individuals were observed cracking nuts by placing them on a stone anvil and hitting them with another large stone (hammer). Similar hammer-and-anvil use has been observed in other wild capuchins including robust capuchin monkeys (genus *Sapajus*) It may take a capuchin up to 8 years to master this skill. The monkeys often transport hard fruits, stones, nuts and even oysters to an anvil for this purpose. Capuchins also use stones as digging tools for probing the substrate and sometimes for excavating tubers. Wild black-striped capuchin use sticks to flush prey from inside rock crevices. Robust capuchins are also known at times to rub defensive secretions from arthropods over their bodies before eating them; such secretions are believed to act as natural insecticides.

Darwin mentioned tool use by wild baboons in *The Descent of Man*:

Brehm states, on the authority of the well-known traveller Schimper, that in Abyssinia when the baboons belonging to one species (*C. gelada*) descend in troops from the mountains to plunder the fields, they sometimes encounter troops of another species (*C. hamadryas*), and then a fight ensues. The Geladas roll down great stones, which the Hamadryas try to avoid.

Darwin continued:

Brehm, when accompanying the Duke of Coburg-Gotha, aided in an attack with fire-arms on a troop of baboons in the pass of Mensa in Abyssinia. The baboons in return rolled so many stones down the mountain, some as large as a man's head, that the attackers had to beat a hasty retreat; and the pass was actually for a time closed against the caravan.

These rather anecdotal reports of stone throwing by baboons have been corroborated by more recent research on chacma baboon (*Papio ursinus*) troops living on the desert floor of the Kuiseb Canyon in South West Africa. Stoning by these baboons is done from the rocky walls of the canyon where they sleep and retreat when they are threatened. Stones are lifted with one hand and dropped over the side. The stones tumble down the side of the cliff or fall directly to the canyon floor. The researchers recorded 23 such incidents involving the voluntary release of 124 stones.

A subadult male from a captive group of Guinea baboons (*Papio papio*) learned, by trial-and-error, to use a tool to rake in food. He then used the tool 104 times over 26 days, thereby providing the group with most of its food.

In April 2018, after four captive baboons managed to escape from their enclosure at Texas Biomedical Research Institute, a 55-gallon barrel left on its side in the pen as an enrichment device was found to have been stood erect next to the perimeter wall, enabling its use as a jumping platform to escape.

In the wild, mandrills have been observed to clean their ears with modified tools. Scientists filmed a large male mandrill at Chester Zoo stripping down a twig, apparently to make it narrower, and then using the modified stick to scrape dirt from underneath its toenails.

In Thailand and Myanmar, crab-eating macaques use stone tools to open nuts, oysters and other bivalves, and various types of sea snails (nerites, muricids, trochids, etc.) along the Andaman sea coast and offshore islands. A troop of wild macaques which regularly interact with humans have learnt to remove hairs from the human's heads, and use the hair to floss their teeth.

Elephants

Elephants show an ability to manufacture and use tools with their trunk and feet. Both wild and captive Asian elephants (*Elephas maximus*) use branches to swat flies or scratch themselves. Eight of 13 captive Asian elephants, maintained under a naturalistic environment, modified branches and switched with the altered branch, indicating this species is capable of the more rare behaviour of tool manufacture. There were different styles of modification of the branches, the most common of which was holding the main stem with the front foot and pulling off a side branch or distal end with the trunk. Elephants have been observed digging holes to drink water, then ripping bark from a tree, chewing it into the shape of a ball thereby manufacturing a "plug" to fill in the hole, and covering it with sand to avoid evaporation. They would later go back to the spot to drink.

Asian elephants may use tools in insightful problem solving. A captive male was observed moving a box to a position where it could be stood upon to reach food that had been deliberately hung out of reach.

Elephants have also been known to drop large rocks onto an electric fence to either ruin the fence or cut off the electricity.

Cetaceans

A community of Indo-Pacific bottlenose dolphins (*Tursiops* sp.) in Shark Bay, Western Australia, made up of approximately 41-54 animals, are known to use conical sponges (*Echinodictyum mesenterinum*) as tools while foraging. This behavior, termed "sponging", occurs when a dolphin breaks off a sponge and wears it over its rostrum while foraging on the seafloor. Sponging behavior typically begins in the second year of life. During sponging, dolphins mainly target fish that lack swim bladders and burrow in the substrate. Therefore, the sponge may be used to protect their rostrums as they forage in a niche where echolocation and vision are less effective hunting techniques. Dolphins tend to carry the same sponge for multiple surfacings but sometimes change sponges. Spongers typically are more solitary, take deeper dives, and spend more time foraging than non-spongers. Despite these costs, spongers have similar calving success to non-spongers.

There is evidence that both ecological and cultural factors predict which dolphins use sponges as tools. Sponging occurs more frequently in areas with higher distribution of sponges, which tends to occur in deeper water channels. Sponging is heavily sex-biased to females. Genetic analyses suggest that all spongers are descendants of a single matriline, suggesting cultural transmission of the use of sponges as tools. Sponging may be socially learned from mother to offspring. Social grouping behavior suggests homophily (the tendency to associate with similar others) among dolphins that share socially learned skills such as sponge tool use. Sponging has only been observed in Shark Bay.

Indo-Pacific bottlenose dolphins in Shark Bay have also been observed carrying conch shells. In this behavior, dolphins insert their rostrum into the shell's aperture. Although this behavior is rare, it appears to be used for foraging. Dolphins appear to use the conch shells to scoop fish from the substrate then carry the shell to retrieve the fish near the surface.

Sea Otters

A sea otter using a rock to break open a shell

Under each foreleg, the sea otter (*Enhydra lutris*) has a loose pouch of skin that extends across the chest. In this pouch (preferentially the left side), the animal stores collected food to bring to the surface. This pouch also holds a rock, unique to the otter, that is used to break open shellfish and clams. To open hard shells, it may pound its prey with both paws against the rock which it places on its chest. Furthermore, sea otters will use large stones to pry an abalone off its rock; they will hammer the abalone shell with observed rates of 45 blows in 15 seconds or 180 rpm, and do it in two or three dives. Releasing an abalone, which can cling to rock with a force equal to 4,000 times its own body weight, requires multiple dives by the otter.

Other Carnivores

Wild banded mongooses (*Mungos mungo*) regularly use anvils to open food items with a hard shell such as rhinoceros beetles, bird eggs, snail shells or pupating dung beetles. They use a range of anvils commonly including rocks and the stems of trees, but will also use the side-walls of gullys and even dried elephant dung. Pups as young as 2 months of age are already showing the behavioural patterns associated with using an anvil, however, successful smashing is usually shown in individuals older than 6 months of age.

Honey badgers both wild and captive have been filmed manipulating various objects to assist them in making climbs.

North American badgers (*Taxidea taxus*) hunt Richardson's ground squirrels (*Spermophilus richardsonii*). The most common hunting technique is excavation of burrow systems, but plugging of openings into ground-squirrel tunnels accounts for 5–23% of hunting actions. Badgers usually use soil from around the tunnel opening, or soil dragged 30–270 cm from a nearby mound to plug tunnels. The least common (6%), but most novel, form of plugging used by 1 badger involved movement of 37 objects from distances of 20–105 cm to plug openings into 23 ground-squirrel tunnels on 14 nights.

In 2011, researchers at the Dingo Discovery and Research Centre in Melbourne, Australia, filmed a dingo manipulating a table and using this to get food.

Moulting brown bears in Alaska have been observed using rocks to exfoliate.

In Birds

Tool use is found in at least thirty-three different families of birds. According to Jones and Kamil's definition, a bearded vulture dropping a bone on a rock would not be considered using a tool since the rock cannot be seen as an extension of the body. However, the use of a rock manipulated using the beak to crack an ostrich egg would qualify the Egyptian vulture as a tool user. Many other species, including parrots, corvids and a range of passerines, have been noted as tool users.

Bird Nests Show a Great Diversity in Complexity

Intricate nests of weaver birds

Relatively simple nest of a greater black-backed gull

Emperor penguins do not build a nest

Many birds (and other animals) build nests. It can be argued that this behaviour constitutes tool use according to the definitions given above; the birds "carry objects (twigs, leaves) for future use", the shape of the formed nest prevents the eggs from rolling away and thereby "extends the physical influence realized by the animal", and the twigs are bent and twisted to shape the nest, i.e. "modified to fit a purpose". The complexity of bird nests varies markedly, perhaps indicating a range in the sophistication of tool use. For example, compare the highly complex structures of weaver birds to the simple mats of herbaceous matter with a central cup constructed by gulls, and it is noteworthy that some birds do not build nests, e.g. emperor penguins. The classification of nests as tools has been disputed on the basis that the completed nest, or burrow, is not held or manipulated.

Woodpecker Finches

Perhaps the best known and most studied example of an avian tool user is the woodpecker finch (*Camarhynchus pallidus*) from the Galápagos Islands. If the bird uncovers prey in bark which is inaccessible, the bird then flies off to fetch a cactus spine which it may use in one of three different ways: as a goad to drive out an active insect (without necessarily touching it); as a spear with which to impale a slow-moving larva or similar animal; or as an implement with which to push, bring towards, nudge or otherwise manoeuvre an inactive insect from a crevice or hole. Tools that do not exactly fit the purpose are worked by the bird and adapted for the function, thus making the finch a "tool maker" as well as a "tool user". Some individuals have been observed to use a different type of tool with novel functional features such as barbed twigs from blackberry bushes, a plant that is not native to the islands. The twigs were first modified by removing side twigs and leaves and then used such that the barbs helped drag prey out of tree crevices.

There is a genetic predisposition for tool use in this species, which is then refined by individual trial-and-error learning during a sensitive phase early in development. This means that, rather than following a stereotypical behavioural pattern, tool use can be modified and adapted by learning.

The importance of tool use by woodpecker finch species differs between vegetation zones. In the arid zone, where food is limited and hard to access, tool use is essential, especially during the dry season. Up to half of the finches' prey is acquired with the help of tools, making them even more routine tool users than chimpanzees. The tools allow them to extract large, nutritious insect larvae from tree holes, making tool use more profitable than other foraging techniques. In contrast, in the humid zone, woodpecker finches rarely use tools, since food availability is high and prey is more easily obtainable. Here, the time and energy costs of tool use would be too high.

There have been reported cases of woodpecker finches brandishing a twig as a weapon.

Corvids

Corvids are a family of birds characterised by relatively large brains, remarkable behavioural plasticity (especially highly innovative foraging behaviour) and well-developed cognitive abilities.

New Caledonian Crow

New Caledonian crows (*Corvus moneduloides*) are perhaps the most studied corvid with respect to tool-use.

In the wild, they have been observed using sticks as tools to extract insects from tree bark. The birds poke the insects or larvae until they bite the stick in defence and can then be drawn out. This "larva fishing" is very similar to the "termite fishing" practised by chimpanzees. In the wild, they also manufacture tools from twigs, grass stems or similar plant structures, whereas captive individuals have been observed to use a variety of materials, including feathers and garden wire. Stick tools can either be non-hooked—being more or less straight and requiring only little modification—or hooked. Construction of the more complex hooked tools typically involves choosing a forked twig from which parts are removed and the remaining end is sculpted and sharpened. New Caledonian crows also use pandanus tools, made from barbed leaf edges of screw pines (*Pandanus* spp.) by precise ripping and cutting although the function of the pandanus tools is not understood.

While young birds in the wild normally learn to make stick tools from elders, a laboratory New Caledonian crow named "Betty" was filmed spontaneously improvising a hooked tool from a wire. It was known that this individual had no prior experience as she had been hand-reared. New Caledonian crows have been observed to use an easily available small tool to get a less easily available longer tool, and then use this to get an otherwise inaccessible longer tool to get food that was out of reach of the shorter tools. One bird, "Sam", spent 110 seconds inspecting the apparatus before completing each of the steps without any mistakes. This is an example of sequential tool use, which represents a higher cognitive function compared to many other forms of tool use and is the first time this has been observed in non-trained animals. Tool use has been observed in a non-foraging context, providing the first report of multi-context tool use in birds. Captive New Caledonian crows have used stick tools to make first contact with objects that were novel and hence potentially dangerous, while other individuals have been observed using a tool when food was within reach but placed next to a model snake. It has been claimed "Their [New Caledonian crow] tool-making skills exceed those of chimpanzees and are more similar to human tool manufacture than those of any other animal."

New Caledonian crows have also been observed performing tool use behaviour that had hitherto not been described in non-human animals. The behaviour is termed "insert-and-transport tool use". This involves the crow inserting a stick into an object and then walking or flying away holding both the tool and object on the tool.

Hawaiian Crow

Captive individuals of the critically endangered Hawaiian crow (*Corvus hawaiiensis*) use tools to extract food from holes drilled in logs. The juveniles exhibit tool use without training or social learning from adults. As 104 of the 109 surviving members of the species were tested, it is believed to be a species-wide ability.

Others

Other corvid species, such as rooks (*Corvus frugilegus*), can also make and use tools in the laboratory, showing a degree of sophistication similar to that of New Caledonian crows. While not confirmed to have used tools in the wild, captive blue jays (*Cyanocitta cristata*) have been observed using strips of newspaper as tools to obtain food.

Various corvids have reached for stones to place in a vessel of water so as to raise the surface level to drink from it or access a floating treat, enacting Aesop's Fable of The Crow and the Pitcher.

A wild American crow (*Corvus brachyrhynchos*) has been observed to modify and use a piece of wood as a probe. Green jays (*Cyanocorax yncas*) have been observed using sticks as tools to extract insects from tree bark. Large-billed crows in urban Japan have been filmed using an innovative technique to crack hard-shelled nuts by dropping them onto crosswalks (pedestrian crossings) and letting them be run over and cracked by cars. They then retrieve the cracked nuts when the cars are stopped at the red light. In some towns in America, crows drop walnuts onto busy streets so that the cars will crack the nuts. Hooded crows (*Corvus cornix*) use bait to catch fish. Individuals (who may have observed fish being fed bread by humans) will place the bread in the water to attract fish.

Common ravens (*Corvus corax*) are one of only a few species who make their own toys. They have been observed breaking off twigs to play with socially. A corvid has been filmed sliding repeatedly down a snow-covered roof while balancing on a lid or tray. Another incidence of play in birds has been filmed showing a corvid playing with a table tennis ball in partnership with a dog, a rare example of tool use for the purposes of play. Blue jays, like other corvids, are highly curious and are considered intelligent birds. Young blue jays playfully snatch brightly coloured or reflective objects, such as bottle caps or pieces of aluminium foil, and carry them around until they lose interest.

Warblers

A pouch containing a tailorbird's nest

The tailorbird (genus *Orthotomus*) takes a large growing leaf (or two or more small ones) and with its sharp bill pierces holes into opposite edges. It then grasps spider silk, silk from cocoons, or plant fibres with its bill, pulls this "thread" through the two holes, and knots it to prevent it from pulling through (although the use of knots is disputed). This process is repeated several times until the leaf or leaves forms a pouch or cup in which the bird then builds its nest. The leaves are sewn together in such a way that the upper surfaces are outwards making the structure difficult to see. The punctures made on the edge of the leaves are minute and do not cause browning of the leaves, further aiding camouflage. The processes used by the tailorbird have been classified as sewing, rivetting, lacing and matting. Once the stitch is made, the fibres fluff out on the outside and in effect they are more like rivets. Sometimes the fibres from one rivet are extended into an adjoining puncture and appear more like sewing. There are many variations in the nest and some may altogether lack the cradle of leaves. It is believed that only the female performs this sewing behaviour. The Latin binomial name of the common tailorbird, *Orthotomus sutorius*, means "straight-edged" "cobbler" rather than tailor. Some birds of the genus *Prinia* also practice this sewing and stitching behaviour.

Brown-headed Nuthatches

Brown-headed nuthatches (*Sitta pusilla*) have been observed to methodically use bark pieces to remove other flakes of bark from a tree. The birds insert the bark piece underneath an attached bark scale, using it like a wedge and lever, to expose hiding insects. Occasionally, they reuse the same piece of bark several times and sometimes even fly short distances carrying the bark flake in their beak. The evolutionary origin of this tool use might be related to these birds frequently wedging seeds into cracks in the bark to hammer them open with their beak, which can lead to bark coming off.

Brown-headed nuthatches have used a bark flake to conceal a seed cache.

Crested Lark

A Crested lark (*Galerida cristata*) has been photographed apparently holding in its bill a stone chip it was reportedly using to dislodge prey from paving joints.

Parrots

Kea, a highly inquisitive New Zealand mountain parrot, have been filmed stripping twigs and inserting them into gaps in box-like stoat traps to trigger them. Apparently, the kea's only reward is the banging sound of the trap being set off. In a similarly rare example of tool preparation, a captive Tanimbar corella (*Cacatua goffiniana*) was observed breaking off and "shaping" splinters of wood and small sticks to create rakes that were then used to retrieve otherwise unavailable food items on the other side of the aviary mesh. This behaviour has been filmed.

Many owners of household parrots have observed their pets using various tools to scratch various parts of their bodies. These tools include discarded feathers, bottle caps, popsicle sticks, matchsticks, cigarette packets and nuts in their shells.

Hyacinth macaws (*Anodorhynchus hyacinthinus*) have been repeatedly observed to use tools when breaking open nuts, for example, pieces of wood being used as a wedge. Several birds have wrapped a piece of leaf around a nut to hold it in place. This behaviour is also shown by palm cockatoos (*Probosciger aterrimus*). It seems that the hyacinth macaw has an innate tendency to use tools during manipulation of nuts, as naïve juveniles tried out a variety of objects in combination with nuts.

Egyptian Vultures

When an Egyptian vulture (*Neophron percnopterus*) encounters a large egg, it takes a stone into its beak and forcefully throws it at the egg until the shell is broken, usually taking a few minutes. This behaviour, first reported in 1966, seems to be largely innate and is displayed by naïve individuals. Its origin could be related to the throwing of eggs; rounded (egg-like) stones are preferred to jagged ones.

In a small population in Bulgaria, Egyptian vultures use twigs to collect sheep wool for padding their nests. Although both twigs and wool can serve as nesting material, this appears to be deliberate tool use. The birds approached bits of discarded wool with a twig in their beak, which was then either used as a rake, to gather the wool into heaps, or to roll up the wool. Wool was collected only after shearing or simulated shearing of sheep had taken place, but not after wool had simply been deposited in sheep enclosures.

Fire-foraging Raptors

In Australia the black kite (*Milvus migrans*), whistling kite (*Haliastur sphenurus*) and unrelated brown falcon (*Falco berigora*) are not only attracted to wildfires to source food, but will variously use their beaks or talons to carry burning sticks so as to spread fire, complicating human efforts to contain fires using firebreaks.

Owls

Burrowing owls (*Athene cunicularia*) frequently collect mammalian dung, which they use as a bait to attract dung beetles, a major item of prey.

Gulls

Seagulls have been known to drop live oyster shells on paved and hard surfaces so that cars can drive over them and break the shell. So many get dropped that it is difficult to drive safely near some waterways. Certain species (e.g. the herring gull) have exhibited tool use behavior, using pieces of bread as bait to catch goldfish.

Herons

The green heron (*Butorides virescens*) and its sister species the striated heron (*Butorides striata*) have been recorded using food (bread crusts), insects, leaves, and other small objects as bait to attract fish, which they then capture and eat.

In Reptiles

Tool use by American alligators and mugger crocodiles has been documented. During the breeding season, birds such as herons and egrets look for sticks to build their nests. Alligators and crocodiles collect sticks to use as bait to catch birds. The crocodilian positions itself near a rookery, partially submerges with the sticks balanced on its head, and when a bird approaches to take the stick, it springs its trap. This stick displaying strategy is the first known case of a predator not only using an object as a lure, but also taking into account the seasonal behavior of its prey.

In Fish

Several species of wrasses have been observed using rocks as anvils to crack bivalve (scallops, urchins and clams) shells. It was first filmed in an orange-dotted tuskfish (*Choerodon anchorago*) in 2009 by Giacomo Bernardi. The fish fans sand to unearth the bivalve, takes it into its mouth, swims several metres to a rock which it uses as an anvil by smashing the mollusc apart with sideward thrashes of the head. This behaviour has been recorded in a blackspot tuskfish (*Choerodon schoenleinii*) on Australia's Great Barrier Reef, yellowhead wrasse (*Halichoeres garnoti*) in Florida and a six-bar wrasse (*Thalassoma hardwicke*) in an aquarium setting. These species are at opposite ends of the phylogenetic tree in this family, so this behaviour may be a deep-seated trait in all wrasses.

It has been reported that freshwater stingrays use water as a tool by manipulating their bodies to direct a flow of water and extract food trapped amongst plants.

Prior to laying their eggs on a vertical rock face, male and female whitetail major damselfish clean the site by sand-blasting it. The fish pick up sand in their mouths and spit it against the rock face. Then they fan the area with their fins. Finally they remove the sand grains that remain stuck to the rock face by picking them off with their mouths.

Banded acara, (*Bujurquina vittata*), South American cichlids, lay their eggs on a loose leaf. The male and female of a mating pair often "test" leaves before spawning: they pull and lift and turn candidate leaves, possibly trying to select leaves that are easy to move. After spawning, both parents guard the eggs. When disturbed, the parent acara often seize one end of the egg-carrying leaf in their mouth and drag it to deeper and safer locations.

Archerfish are found in the tropical mangrove swamps of India and Australasia. They approach the surface, take aim at insects that sit on plants above the surface, squirt a jet of water at them, and grab them after the insects have been knocked off into the water. The jet of water is formed by the action of the tongue, which presses against a groove in the roof of the mouth. Some archerfish can hit insects up to 1.5 m above the water surface. They use more water, which gives more force to the impact, when aiming at larger prey. Triggerfish (*Pseudobalistes fuscus*) blow water to turn sea urchins over and expose their more vulnerable ventral side. Whether these later examples can be classified as tool use depends on which definition is being followed because there is no intermediate or manipulated object, however, they are examples of highly specialized natural adaptations.

In Invertebrates

Cephalopods

A small octopus (4-5 cm, c. 2-inch diameter) using a nut shell and clam shell as shelter

At least four veined octopus (*Amphioctopus marginatus*) individuals were witnessed retrieving coconut shells, manipulating them, stacking them, transporting them some distance (up to 20 metres), and then reassembling them to use as a shelter. The octopuses use coconut shells discarded by humans which have eventually settled in the ocean. They probe their arms down to loosen the mud, then rotate the shells out. After turning the shells so the open side faces upwards, the octopuses blow jets of mud out of the bowl before extending their arms around the shell—or if they have two halves, stacking them first, one inside the other. They then stiffen their legs and move away in a manner which has been called "stilt-walking". The octopuses eventually use the shells as a protective shelter in areas where little other shelter exists. If they just have one half, they simply turn it over and hide underneath. But if they are lucky enough to have retrieved two halves, they assemble them back into the original closed coconut form and sneak inside. This behaviour has been filmed. The authors of the research article claimed this behaviour falls under the definition of tool use because the shells are carried for later use. However, this argument remains contested by a number of other biologists who state that the shells actually provide continuous protection from abundant bottom-dwelling predators in their home range.

Octopuses deliberately place stones, shells and even bits of broken bottle to form a wall that constricts the aperture to the den, a type of tool use.

In laboratory studies, *Octopus mercatoris*, a small pygmy species of octopus, has been observed to block its lair using a plastic Lego brick.

Smaller individuals of the common blanket octopus (*Tremoctopus violaceus*) hold the tentacles of the Portuguese man o' war, to whose poison they are immune, both as protection and as a method of capturing prey.

Insects

Ants of the species *Conomyrma bicolor* pick up stones and other small objects with their mandibles and drop them down the vertical entrances of rival colonies, allowing workers to forage for food without competition.

Several species of ant are known to use substrate debris such as mud and leaves to transport water to their nest. A study in 2017 reported that when two species of Aphaenogaster ant are offered natural and artificial objects as tools for this activity, they choose items with a good soaking capacity. The ants develop a preference for artificial tools that cannot be found in their natural environment, indicating plasticity in their tool-use behaviour.

Hunting wasps of the genus *Prionyx* use weights (such as compacted sediment or a small pebble) to settle sand surrounding a recently provisioned burrow containing eggs and live prey in order to camouflage and seal the entrance. The wasp vibrates its wing muscles with an audible buzz while holding the weight in its mandibles, and applies the weight to the sand surrounding its burrow, causing the sand to vibrate and settle. Another hunting wasp, *Ammophila*, uses pebbles to close burrow entrances.

Insects can also learn to use tools. A study in 2017 showed that bumblebees of the species *Bombus terrestris* learned to move a small wooden ball to the goal in order to get sucrose reward.

References

- Englar, Ryane E. (2017). Performing the Small Animal Physical Examination. John Wiley & Sons. ISBN 9781119295310
- Gregory, Neville G. (2008). Physiology and Behaviour of Animal Suffering. John Wiley & Sons. p. 107. ISBN 9781405173025
- Matoba, Tomoyuki; Kutsukake, Nobuyuki; Hasegawa, Toshikazu (4 September 2013). "Head Rubbing and Licking Reinforce Social Bonds in a Group of Captive African Lions, Panthera leo". PLOS ONE. 8 (9): e73044. doi:10.1371/journal.pone.0073044. ISSN 1932-6203
- Andrew right, A Criticism of the IASP's Definition of Pain, "Archived copy". Archived from the original on 22 August 2016. Retrieved 01, February 2020
- Fendt, Markus; Kiyokawa, Yasushi; Endres, Thomas (2016). Scents that matter - from olfactory stimuli to genes, behaviors and beyond. Frontiers Media SA. p. 34. ISBN 9782889198139
- Leake, J. (14 March 2004). "Anglers to Face RSPCA Check". The Sunday Times. Archived from the original on 23 September 2015. Retrieved 21, April 2020
- Kim, E.J.; Kim, E.S.; Covey, E.; Kim, J.J. (2010). "Social transmission of fear in rats: the role of 22-kHz ultrasonic distress vocalization". PLoS ONE. 5 (12): e15077. doi:10.1371/journal.pone.0015077. PMC 2995742 . PM ID 21152023
- Schaller, George B. (2009). The Serengeti Lion: A Study of Predator-Prey Relations. University of Chicago Press. ISBN 9780226736600. Retrieved 09, June 2020
- Ramachandran, V.S. (1 January 2009). "Self Awareness: The Last Frontier, Edge Foundation web essay". Retrieved 25, March 2020

Permissions

All chapters in this book are published with permission under the Creative Commons Attribution Share Alike License or equivalent. Every chapter published in this book has been scrutinized by our experts. Their significance has been extensively debated. The topics covered herein carry significant information for a comprehensive understanding. They may even be implemented as practical applications or may be referred to as a beginning point for further studies.

We would like to thank the editorial team for lending their expertise to make the book truly unique. They have played a crucial role in the development of this book. Without their invaluable contributions this book wouldn't have been possible. They have made vital efforts to compile up to date information on the varied aspects of this subject to make this book a valuable addition to the collection of many professionals and students.

This book was conceptualized with the vision of imparting up-to-date and integrated information in this field. To ensure the same, a matchless editorial board was set up. Every individual on the board went through rigorous rounds of assessment to prove their worth. After which they invested a large part of their time researching and compiling the most relevant data for our readers.

The editorial board has been involved in producing this book since its inception. They have spent rigorous hours researching and exploring the diverse topics which have resulted in the successful publishing of this book. They have passed on their knowledge of decades through this book. To expedite this challenging task, the publisher supported the team at every step. A small team of assistant editors was also appointed to further simplify the editing procedure and attain best results for the readers.

Apart from the editorial board, the designing team has also invested a significant amount of their time in understanding the subject and creating the most relevant covers. They scrutinized every image to scout for the most suitable representation of the subject and create an appropriate cover for the book.

The publishing team has been an ardent support to the editorial, designing and production team. Their endless efforts to recruit the best for this project, has resulted in the accomplishment of this book. They are a veteran in the field of academics and their pool of knowledge is as vast as their experience in printing. Their expertise and guidance has proved useful at every step. Their uncompromising quality standards have made this book an exceptional effort. Their encouragement from time to time has been an inspiration for everyone.

The publisher and the editorial board hope that this book will prove to be a valuable piece of knowledge for students, practitioners and scholars across the globe.

Index

Lightning Source UK Ltd.
Milton Keynes UK
UKHW051612210922
409191UK00002B/99

9 781639 895779